When complexity means failure...

ERECTUS

Why is it the ‘wrong’ species that survived?

editorial@ciencias.bio.br

International Cataloging-in-Publication Data (CIP)

Heemann, Felipe.

Erectus: Why is it the 'wrong' species that survived? / Felipe Heemann — 1st ed. — São Paulo, SP: Brain Codex, 2026.

Bibliography
ISBN 978-65-988126-6-9

1. Human evolution 2. Evolutionary biology 3. Paleoanthropology 4. Philosophy of biology 5. Cognitive neuroscience 6. Symbolic cognition 7. Hypermodernity 8. Naturalism I. Title.

CDD – 599.93

Index for systematic catalog
1. Human evolution — Philosophy

When complexity means failure...

ERECTUS

Why is it the 'wrong' species that survived?

FELIPE HEEMANN A. RIBEIRO

Professor, evolutionary biologist, molecular neurobiologist, and linguist — with a scholarly formation that extends equally into Physics and Mathematics — professor Heemann has dedicated his intellectual trajectory to reconstructing the bridges between Science and the Humanities that modernity has systematically dismantled. A graduate in Biological Sciences from the University of São Paulo (USP), his scholarship moves with ease across these fields and into Philosophy, tracing the contours of a restless and transdisciplinary mind. His work goes beyond the mechanisms of life to interrogate the transformations that scientific knowledge provokes in how we think and inhabit the world. His essays invite the reader into the history of ideas — guided by scientific rigor and the sensitivity of a committed educator. In his pages, Science ceases to be mere explanation: it becomes narrative, beauty, and revelation.

BRAIN CODEH
NEUROBIOLOGIA

ISBN: 978-3-12732-320-7

9 783127 323207

I dedicate these writings to my son.

If these pages trace the wreckage of our species' pride, your existence is the one shore where I still choose to land.

In a world so vastly and beautifully wrong, you are my only right answer — the living proof that even here, even now, even in a civilization losing its way, love remains the one structure that holds.

In memoriam

Melzinha.
A life of companionship, tenderness,
shy silences, and the occasional fierce protest.

Knowledge has two extremes which meet.

The first is the natural ignorance in which all men are born.

The other is that reached by great minds, who, having run through all that men can know, find that they know nothing, and come back to the same ignorance from which they set out.

But it is a learned ignorance which knows itself.

Those in between, who have left this natural ignorance and have not reached the other, have some smattering of knowledge, and pretend to be wise.

These disturb the world and judge everything badly.

The common people and the truly wise form the order of the world; the latter despise it and are despised.

BLAISE PASCAL (1623-1662)
Pensées

矛盾

PREFACE

This book starts from an uncomfortable premise: that the species which endured may not have been the one best suited to do so. Not as metaphor or rhetorical provocation, but as a conclusion grounded in factual evidence that we rarely examine with the honesty it deserves. We celebrate ***Homo sapiens*** as the pinnacle of evolution — builder of cities, creator of art, decoder of the cosmos — as though such a pinnacle existed at all. We measure our success by the sophistication of our technologies, the complexity of our philosophies, the beauty of our symphonies. But these are criteria we ourselves established: narcissistic mirrors reflecting our own values, not the indifference of Nature.

If, instead, we measure evolutionary success by the criteria that Nature itself appears to impose — longevity in deep time, sustainable integration with the environment, absence of self-imposed threats — we discover something profoundly unsettling. ***Homo erectus***, the one we deemed primitive and obsolete, lived for approximately 1.5 to 1.9 million years in relative equilibrium with the Biosphere. It traversed glacial and interglacial ages, adapted to diverse environments from Africa to Asia, maintained a functional lithic technology — the Acheulean — for hundreds of thousands of years. And it existed, in the deepest sense of that word, simply existed, generation after generation, for a span of time our minds can barely conceive.

We, ***Homo sapiens***, have existed for approximately 300,000 years. Of those, only the last ten millennia include what we call civilization. And in those ten thousand years, we built nuclear arsenals capable of extinguishing our species, destabilized the planetary climate through the relentless extraction of resources, and created inequalities on a scale that allows roughly 10% of the population to control 80% of the world's resources. We have produced an epidemic of psychological suffering — anxiety, depression, existential emptiness — at levels that would have been unimaginable to our ancestors. The numbers do not lie: measured by natural criteria of success, we are not winning. We are, at best, a risky experiment whose outcome remains uncertain.

The central distinction this book proposes is between efficacy and effectiveness. Efficacy is the individual or collective capacity to solve specific problems, to innovate, to create solutions for immediate challenges. In this regard, sapiens is extraordinary. We developed Agriculture, Writing, Mathematics, Quantum Physics, Medicine, Computing. We placed human artifacts on other planets. We mapped our own genome. Our efficacy in solving technical problems is undeniable.

But effectiveness is something else entirely. Effectiveness is the sustainable survival of a species in deep time — the capacity to persist through environmental change without creating the conditions for its own extinction. Effectiveness is not a value judgment; it is a descriptive, non-normative measure of persistence across geological time — what Van Valen (1973) formalized as the macroevolutionary currency when he demonstrated that extinction probability remains roughly constant per unit time within taxonomic groups, regardless of how long a lineage has already survived. By that metric, duration is the only scoreboard Nature keeps. And here, sapiens reveals itself to be grotesquely ineffective. Every advance in our efficacy seems to generate new risks to our effectiveness. The same intelligence that cures disease created chemical and biological weapons. The same capacity for planning that built cities generated ecological collapse. The same language that enables Philosophy also enables the construction of abstract categories of 'others' who can be exterminated without empathy.

And it is precisely here that the argument finds its most uncomfortable core — the place where neuroscience and philosophy meet in ways we would prefer to avoid.

Erectus had a sophisticated nervous system. It was not a creature without anticipation, without memory, without response to danger. Like any vertebrate, it processed the imminence of threat, integrating immediate past and near future within a continuous sensorimotor window that is the very texture of action. A predator anticipating its leap is not "thinking about the future": it is completing a pattern already underway in the body. This is the temporality of organisms without recursive language — not a contemplative achievement, not a romanticized primitive wisdom, but what we might call the temporality of imminence: the mode in which the nervous system processes the world as a continuous field of action. Body and world coincide. There is no additional layer between them. "Present" here is not a phenomenological category, but a neurocognitive horizon defined by short-range prediction and limited symbolic recursion.

What erectus almost certainly did not have was something else entirely: the capacity to construct symbolic castles and inhabit them in the complete absence of their referent. Neanderthals, who shared part of our symbolic and cognitive architecture, likely possessed intermediate forms of temporal projection; erectus did not. The anxiety that corrodes sapiens is not a response to the present predator — it is a response to the predator that recursive language keeps alive and affectively present decades after it has vanished, or projects with terrifying clarity into futures that do not yet exist and may never exist. The limbic system responds to a real threat and a narrated one with extensively overlapping activation patterns — the amygdala, in particular, recruits substantially similar circuits for both (Phelps & LeDoux, 2005; Schiller *et al.*, 2009). The distinction exists at the cortical level, but at the subcortical level where fear is generated, both are treated as presence. We are the only species known to suffer systematically from worlds that do not exist — worlds constructed through recursive language rather than immediate threat — that linguistically inhabits a past that no longer exists and a future that has not yet arrived, while the body occupies a present the mind frequently ignores.

This is not metaphor. It is the direct neurological consequence of a language that became recursive and symbolic to the point of constructing inhabitable parallel worlds. Erectus did not need contemplative practices to "return" to the present because it probably, as far as current evidence allows us to infer, never possessed the symbolic scaffolding that allows us to absent ourselves from it so efficiently. It was not wiser. It had a different cognitive architecture — and that difference, which we celebrate as superiority, carries a cost we rarely account for.

The argument developed across six sections is that this disparity between efficacy and effectiveness is not accidental, but the result of cognitive hypertrophy. The sapiens brain expanded — particularly in the prefrontal and parietal cortices — in ways that generated extraordinary cognitive capacities. But this was not necessarily a linear progression toward superior adaptation — Nature, after all, has no register for either progress or superiority. These are categories we projected onto a process that operates exclusively by persistence and reproductive fit, indifferent to the complexity or elegance of its outcomes. It was a development that produced both wonders and curses, both art and genocide, both philosophy and nuclear threat. Neural hypertrophy gave us powerful cognitive tools, but also condemned us to forms of suffering that erectus almost certainly never experienced, because they require symbolic recursion, autobiographical memory, and long-range temporal simulation — not because it was happier by nature, but because it lacked the symbolic instruments necessary to build the prisons we inhabit.

The ideas gathered here were born during a period of forced margin — two years in which the productive system dispensed with me and made me survive through what we will argue exists but is systematically supplanted: the genuine compassion of others. That involuntary margin allowed something rare in hypermodernity: time directed toward thinking without immediate utilitarian purpose. And thinking led to questioning assumptions we rarely examine — among them, the notion that we are superior to our ancestors, which seems obvious to everyone, but is not.

The tone is essayistic but grounded in evidence. Provocative, certainly, but without ideological invective. The goal is not to vilify humanity or to romanticize a primitive past. It is to examine honestly whether our criteria of success make sense when confronted with natural reality. The world's religions offer comforting narratives about our special place in the Cosmos. Science offers a body of evidence indifferent to those narratives. This book takes the second path — while treating human beliefs, including religious ones, with analytical respect: not with contempt, but as symptom.

It is here that Pascal offers a compass. He distinguished three types of people: those who remain in natural ignorance, those who traverse both ignorances and arrive at wisdom, and those who stopped in the middle — who left natural ignorance but could not reach the other end. And it is these last, Pascal says, who "pretend to be wise" and "confuse the world, and judge things badly." This book does not come from the sapiens arrogance of having all the answers. It comes from the attempt to examine our lineage with the same naturalistic criteria we apply to every other species — recognizing that knowing we do not know is the beginning of wisdom, not its end.

What follows is not merely cultural criticism, but a radical reexamination of which lineage demonstrated greater longevity and sustainable integration — measures that Nature registers without intention or purpose. Nature does not reward abstract intelligence or technological capacity. It rewards, in the only sense that matters, what persists. And by that brutal metric, erectus surpasses us by a crushing margin.

This is an uncomfortable thesis. It confronts assumptions that run through virtually all western Philosophy since the Enlightenment. But discomfort, as we shall see, is precisely the distinctive condition of sapiens — we are the species capable of feeling uncomfortable with our own existence, of questioning whether we should be here at all, of torturing ourselves with questions no other animal has ever formulated. Erectus did not suffer from this existential discomfort because it probably lacked the cognitive apparatus necessary to generate it. And perhaps — just perhaps — that was not a deficiency. Perhaps it was an adaptive advantage we discarded in exchange for capacities whose real value has yet to be determined.

I invite you to accompany me along the paths of this argument — not as a student receiving truths, but as an accomplice examining the crime. We will investigate together the sovereignty of the natural and the hidden cost of our cerebral architecture. We will question whether language — not communication, but our recursive and symbolic language — is tool or wound. And we will finally confront the hypothesis that our hypermodern condition is not a passing crisis, but the terminal symptom of a lineage that learned to deny the present by building castles in which it inhabits what does not exist: the past that does not return and the future that has not yet arrived.

If, at the end, you disagree with the conclusion, you will at least have faced the mirror without looking away. If you agree, you will carry a perspective that makes it difficult to see our species with the same innocence again. In either case, the exercise will have been worthwhile. Truth deserves to be examined — especially when it keeps us awake at night.

It was precisely in that insomnia, during two years of forced margin, that these reflections took shape. Written not from an ivory tower, but from the raw ground of necessity, sustained by those rare but ever-present others. What I offer now is the map of that exile — an invitation to look at erectus and see not a primitive sketch, but the silent master we lost.

SUMMARY

I.
The sovereignty of Nature
17

II.
The pitiless mirror
23

III.
The spell of words
41

IV.
The lost present
55

V.
Success and its shadows
69

VI.
Light that blinds
81

I.
The sovereignty of Nature

I do not know what I may appear to the world; but to myself I seem to have been only like a boy playing on the seashore, and diverting myself in now and then finding a smoother pebble or a prettier shell than ordinary, whilst the great ocean of truth lay all undiscovered before me.

Attributed to ISAAC NEWTON
(via Brewster, 1855)

If Newton — who unveiled gravitation and forged the calculus — still saw himself as a child playing on the shore while the ocean of truth lay undiscovered before him, what remains for us but to acknowledge our own smallness before the vast indifference of the Cosmos?

I hold in my hands a fragment of sandstone collected from the Vila Velha formation in Paraná, Brazil. It is approximately 340 million years old. That dating is not a rough estimate based on intuition or oral tradition, but the result of decades of refinement in radiometric methods — uranium–lead, argon–argon — the measurable decay of isotopes, nuclear processes operating with mathematical regularity indifferent to whoever observes them or to whatever they believe. This piece of rock existed when fish were beginning to colonize dry land, when forests of giant ferns dominated the continents, when mammals were an unthinkable future. The rock does not know I am holding it. It knows nothing of me, of my species, of our philosophies or theologies. It simply exists — silent witness to aeons our minds can barely process.

That indifference is not a metaphor. It is the epistemological ground on which any honest investigation of our place in the Cosmos must stand. Nature is unmoved by our stories. It does not adjust its regularities to accommodate our preferences, does not modify its records to validate our narratives, does not consult our cosmologies before determining what is true. Uranium-238 decays into lead-206 with a half-life of 4.47 billion years regardless of how many prayers we offer or how convinced we are that the Earth is six thousand years old. The atoms composing this rock tell a story in a language that admits no alternative interpretations.

Carl Sagan reminded us that we are made of stardust — not as poetry, but as literal description. Every carbon atom in our bodies was forged in the cores of stars that died long before our Sun existed. The iron in our blood, the calcium in our bones, the oxygen we breathe — everything up to the iron nucleus was synthesized in stellar furnaces and scattered across the cosmos by *supernovae*. We are temporary arrangements of stellar matter that, by an accident of complexity, became capable of contemplating themselves. That consciousness does not elevate us above Nature. It merely makes us a piece of Nature aware of itself — for a cosmic instant.

Modern astronomy should have humbled us beyond recovery. Earth is not the center of the Universe, not of the Solar System, not even of particular significance within our own galaxy. We orbit a mediocre star on an undistinguished periphery of a common galaxy among billions. The Milky Way contains between 100 and 400 billion stars. The observable Universe contains roughly two trillion galaxies. On that scale, our existence is statistically irrelevant. If all of human civilization vanished tomorrow — every war, philosophy, symphony, genocide — the universe would not notice. Stars would continue to be born and die, galaxies would collide, and the Cosmos would proceed in its absolute dynamics as though we had never existed.

Sagan called Earth a "pale blue dot," analyzing a photograph taken by Voyager 1 in 1990 from six billion kilometers away. In that image, our planet appears as a single pale pixel suspended in a ray of light.

As Sagan observed in his reflection on the "pale blue dot," everything we know — every person we have loved, every tyrant we have feared, every religion, every scientific discovery — happened on that invisible point. The photograph should induce humility proportional to our insignificance. It rarely does. Sapiens possesses a remarkable capacity to remain unmoved by evidence that wounds its collective vanity.

Emil Cioran offers a necessary counterweight. Where Sagan found melancholic beauty in our cosmic smallness, Cioran saw only confirmation of our irrelevance. Knowing we are made of stardust saves us from nothing. It does not relieve suffering, resolve conflicts, prevent wars, or cure diseases. The awareness of our cosmic insignificance coexists perfectly with anxiety, depression, tribal hatred, and obscene greed. Astronomical knowledge does not transform us morally because morality does not emerge from cosmic facts. It emerges from neurology, evolutionary history, and cultural contingencies — none of which change when we contemplate our position in the Universe.

Life, on this scale, is a local and temporary anomaly. As far as we know, Earth is the only planet where chemistry organized itself into biology. There may be life elsewhere, statistics suggest it, but we have no evidence. And even if it exists, the vast majority of the Universe remains a hostile void where nothing lives, nothing thinks, nothing feels. The Cosmos was not made for us. It was not made for anything. It simply exists, governed by physical laws that carry no intention or purpose. Gravity does not want planets to orbit stars. Entropy does not desire disorder. These are human metaphors for regularities that remain indifferent to us.

Darwin completed the dethronement that Copernicus began. If Copernicus removed us from the spatial center of the Cosmos, Darwin removed us from the teleological center of life. We are not the pinnacle of creation, not the final objective of directed evolution. We are one among millions of species that emerged through descent with modification, differential mortality, and natural selection — mechanisms blind to any goal. Natural selection does not "aim" to create intelligence or consciousness. It simply favors traits that increase survival and reproduction in specific contexts. If consciousness helps, it persists. If not, it disappears.

Twentieth-century Physics dismantled the last refuge of certainty. Newton still allowed imagining the Cosmos as a deterministic machine. Einstein cracked that image. Heisenberg shattered it. Planck, Bohr, and Born revealed that at the quantum level the universe is probabilistic, not deterministic. The act of observation alters what is observed. There is no absolute certainty, no total predictability. Reality does not adjust to what comforts us.

And yet, faced with this convergence of evidence about our insignificance, most humans insist on supernatural narratives that cast them as protagonists of cosmic dramas. Approximately 85% of the world's population identifies with some religion. These religions are mutually contradictory in their central claims. Not all can be true. The correlation between birthplace and belief reveals their cultural origin: born in Saudi Arabia, likely muslim; in Texas, christian; in India, hindu. This geographical predictability is incompatible with the idea that religions grant access to objective transcendent truths.

This is not a critique of religion, but an illustration of how symbolic narratives diverge from natural evidence — a divergence central to the argument of this book. Scientific knowledge converges across cultures because it depends on empirically verifiable evidence. Religious doctrines diverge because they depend on inherited narratives. One corrects itself; the other persists regardless of contradiction.

This epistemological asymmetry matters. When we ask which human lineage "succeeded," we are not making a theological claim. We are proposing a naturalistic criterion: which lineage persisted longer in sustainable integration with its environment? And here Nature records an answer that is unmoved by our preferences. Erectus: between 1.5 and 1.9 million years. Sapiens: 300,000 years, of which the last 10,000 have already created the potential conditions for self-extinction. Measured by criteria Nature appears to impose — longevity, sustainability, absence of self-inflicted threats — erectus surpasses us by a crushing margin.

Why "criteria Nature appears to impose" rather than "objective natural criteria"? Because even that requires epistemic humility. There is no cosmic manual defining success. What we observe is a pattern: species that persist for millions of years in relative equilibrium tend to be considered successful in naturalistic terms. Species that appear briefly and vanish leave fewer traces and existed for statistically insignificant spans of time.

Extinction is the rule, not the exception. Roughly 99% of all species that have ever existed are extinct. They disappeared through climate change, competition, catastrophe, or simple maladaptation. Most extinctions were not dramatic; they were gradual declines until the last individual died without descendants. Nature does not preserve species out of sentimentality. There is no right to continued existence.

We are not exempt. Our intelligence, technology, and art grant no immunity to the processes that eliminated 99% of previous species. In fact, our intelligence may accelerate our extinction. A species capable of building thermonuclear weapons, destabilizing the climate, and creating obscene inequalities is not necessarily better equipped for long-term survival than one that lived in equilibrium for more than a million years.

What do we measure when we measure "evolutionary success"? Traditionally, sapiens evaluates itself by criteria it invented: intelligence, technology, art, philosophy, science. By these metrics, we obviously win. But these are narcissistic metrics. They reflect our values, not the reality of adaptation.

If instead we measure success by longevity and sustainable integration — criteria we apply to every other species — the picture inverts. Bacteria have existed for 3.5 billion years. Sharks for 450 million. Crocodilians survived the extinction that eliminated non-avian dinosaurs. These species have no art or philosophy. But they have something we lack: proven evolutionary success measured in deep time.

Homo erectus lived for between 1.5 and 1.9 million years. This is not speculation — it is paleontological fact. Erectus crossed glacial cycles, adapted to diverse environments, maintained a functional lithic technology for hundreds of thousands of years, and existed in sustainable populations without engineering its own extinction. It built no civilizations, wrote no books, composed no music. But it persisted for a span of time that makes our existence look like a recent experiment whose outcome remains uncertain.

We, by contrast, have existed for 300,000 years, with only the last 10,000 including anything we call civilization. And in those 10,000 years, we created nuclear arsenals, destabilized the climate, triggered a sixth mass extinction, and built economic systems that concentrate resources in ways incompatible with long-term stability. Measured by natural criteria, we are not winning. We are a risky experiment whose success remains undetermined.

This is the sovereignty of Nature: empirical evidence over symbolic narrative. Longevity over brilliance. Persistence over self-destruction. And by that brutal, unsentimental measure, erectus humiliates us.

That humiliation is not moral. It is factual, inscribed in geological strata that neither lie nor console. The rock I hold has existed for 340 million years — before any ape developed a cortex capable of questioning it. And it will exist for millions after the last human word is spoken. The natural endures. We, possibly, will not. Before that geological silence, our arrogance evaporates.

But the deepest dethronement is not cosmological. It is cognitive — and it begins when we examine what our own brain became.

II.
The pitiless mirror

Man shouldn't be able to see his own face – there's nothing more sinister. Nature gave him the gift of not being able to see it, and of not being able to stare into his own eyes.

Only in the water of rivers and ponds could he look at his face. And the very posture he had to assume was symbolic. He had to bend over, stoop down, to commit the ignominy of beholding himself.

The inventor of the mirror poisoned the human heart.

BERNARDO SOARES

Erectus knew humility: to see himself, he had to bow, to lower his body toward earth and water. Sapiens lifted the mirror to eye level and lost that posture, standing erect before his own apparition. Yet the mirror that stirred Soares to pity is darker still: the excess of cortex that turned a fleeting reflection into a permanent symbolic wound. There is no retreat from it. We call it consciousness; he called it a poison.

Between the skull of *Homo erectus* and that of *Homo sapiens*, roughly 600,000 years of anatomical change separate two radically distinct ways of inhabiting the world. The differences are not subtle: while erectus had a cranial capacity ranging from approximately 700 to 1,250 cm^3 across its geographic and temporal span — from the smaller-brained dmanisi specimens to the robust ngandong crania of late Java (Rightmire, 2004; Antón, 2012; Herries *et al.*, 2020), sapiens expanded that volume to 1,200–1,500 cm^3. But this was not a homogeneous enlargement, as though the brain had simply swollen proportionally. Specific regions hypertrophied, carrying functional consequences that deserve to be called by their proper name: burden, not progress.

The prefrontal cortex and the parietal cortex underwent disproportionate expansion. The prefrontal cortex — the region behind the forehead responsible for planning, impulse inhibition, anticipation of future consequences, and the construction of self-narratives — became excessively developed in sapiens. Erectus possessed a functional prefrontal cortex, sufficient for complex motor coordination, tool manufacture, short-term planning, and the protolinguistic communication inferred from endocasts. But it lacked the neurological machinery required to inhabit multiple temporal dimensions at once: a past reconstructed symbolically, a future imagined recursively, and a present reduced to a narrow corridor between both.

The superior parietal cortex expanded in an equally consequential way. This region integrates sensory information, constructs body maps, and allows an organism to know where its body ends and the world begins even with eyes closed. In erectus, this function was sufficiently developed for spatial navigation, object manipulation, and hand-tool coordination. In sapiens, the parietal cortex did more than map the body in space — it generated the abstract representation of a self as an entity separate from the world, as an object among objects, as a subject capable of observing itself from the outside.

Here lies the decisive shift: the emergence of complex self-representation, sustained by these anatomical expansions. Erectus knew it existed — it possessed primary consciousness, the immediate presence in the here-and-now that allows any complex mammal to navigate its environment. Sapiens developed a secondary, reflexive, metacognitive consciousness: it knows that it knows that it exists. It can think about its own thoughts. It can imagine itself being imagined by others. It can construct models of how others construct models of it. This infinite recursion — "I thinking about you thinking about me thinking about you" — became possible because the expanded cortices of sapiens created neurological fractal loops absent in erectus.

Elaborated theories of mind emerge from this architecture. Erectus likely possessed a basic theory of mind: it understood that others held different perspectives, could infer simple intentions, and coordinate

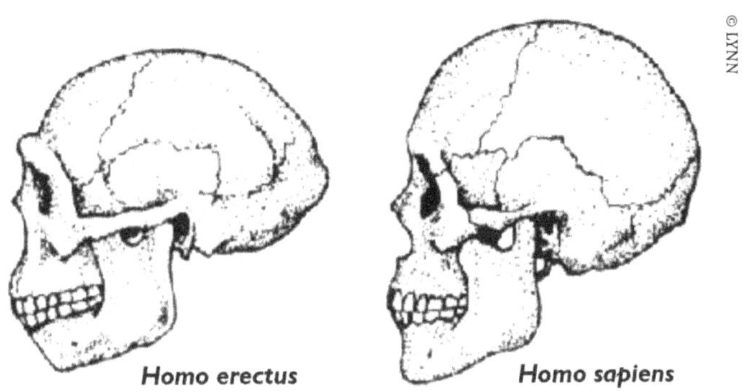

External cranial morphology of *Homo erectus* and *Homo sapiens*. The contrast in vault shape and cranial capacity reflects broader patterns of cortical expansion in sapiens, particularly in parietal, prefrontal, and cingulate regions discussed in the text, even though such internal reorganizations are not directly visible in the external bone surface.

group actions. But sapiens developed third-, fourth-, and fifth-order theory of mind: not only do I know that you know, but I know that you know that I know that you know, and I can anticipate your anticipation of my anticipation. This capacity — technically called representational recursion — is not a neutral cognitive achievement. It is an evolutionary cost. Each additional layer of representation demands more neurons, more synapses, more energy, more cranial space. And for what? To construct increasingly elaborate simulations of worlds that do not exist, of futures that may never materialize, of pasts reconstructed through symbolic narrative rather than adaptive relevance.

The anterior cingulate cortex also expanded significantly in sapiens. This deep structure, involved in emotional regulation, conflict detection, and error monitoring, became hyper-reactive. In erectus, the anterior cingulate functioned as an adaptive alarm system: it signaled inconsistencies between expectation and reality, allowing behavioral adjustments. In sapiens, this system became pathologically sensitive. It detects conflicts where none exist. It anticipates errors that will never occur. It generates chronic anxiety in response to imaginary threats. Obsessive rumination — the tendency to chew the same thought repeatedly without resolution — is plausibly linked to this cingulate hyperactivity.

Then came expanded temporal projection. Erectus lived predominantly within an extended present — a recent past informing immediate decisions, a near future tied to survival. Sapiens developed the capacity to project itself decades forward, to imagine itself old, dying, dead. It can remember events that occurred before it was born through cultural narratives. It can worry about hypothetical scenarios that will never materialize. This capacity for mental time travel — chronesthesia — appears adaptive. Planning the distant future should increase survival chances. But the price is permanent displacement from the only time that exists: the instant that already passed while you finished reading this sentence.

The anatomical changes do not stop at the cortex. Endocasts of erectus — internal molds of the cranial cavity — reveal neural reorganizations consistent with protolanguage: vocal communication richer than simple calls, likely coordinated with gesture. But nothing in the fossil or comparative record suggests recursive grammar, clause embedding, or the hierarchical syntax that allows sentences like the one you are reading now. Recursive language is not a neutral communication tool. It is a cognitive technology that restructures experience itself. When you name an emotion, it ceases to be lived directly and becomes mediated by concept. When you narrate your life as a story with beginning, middle, and end, you transform lived experience into an object of observation. Language creates distance between the one living and the lived. Erectus did not have that distance.

The sapiens hippocampus also shows significant differences. This structure, essential for episodic memory, became more developed, allowing not merely the remembering that something happened, but the mental reliving of the entire context: where one was, what one saw, what one felt, what one thought. Detailed episodic memory seems beneficial — who wouldn't want to remember vividly the best moments of life? But it also means reliving traumas with cinematic precision. It means carrying decades of painful experiences that could have been forgotten. It means constructing a narrative identity anchored in events that no longer carry adaptive relevance.

All these anatomical changes converge to create what we might call the "syndrome of the permanent internal mirror". Erectus looked outward: at prey, at terrain, at the group. Its attention was externalized, directed at the present world. Sapiens developed the capacity to turn attention inward, to observe its own mental processes, monitor its emotions, evaluate its thoughts. This metacognitive self-observation — sustained by expanded cortical areas — creates an infinite loop: "I think, therefore I observe that I think, therefore I think about observing myself thinking."

Modern neuroscience adds a further layer to this internal mirror. The brain does not simply register the world; it predicts it. It uses past experience to generate models of the immediate future and lives inside those projections, correcting them when reality fails to match expectation. Perception is thus the brain's best guess about what is about to happen, not what is happening now. And the self emerges as the most stable of these predictions — the narrative residue of countless attempts to minimize error between what was anticipated and what occurred. A creature capable of recursive language and long-range temporal simulation becomes, inevitably, a creature that inhabits the thin horizon of the almost-future rather than the present.

And here lies the point the triumphalist narrative of cognitive progress ignores: metacognitive recursion is not necessarily adaptive. Thinking about your own thoughts does not improve your ability to find food, avoid predators, or reproduce. It may hinder it — the animal that ruminates about its ruminations while the predator approaches leaves no descendants. Metacognition has costs: energetic, psychological, existential. Anxiety, depression, dissociation, depersonalization — disorders that emerge when self-observation becomes pathological — are disorders specific to the cognitive architecture of ***Homo sapiens***.

Erectus made tools, hunted cooperatively, controlled fire, cared for sick group members, and perhaps — just perhaps — buried its dead with some degree of ritual. It had sophisticated cognition. But it lacked the neuroanatomical machinery to ask "Who am I?", "What is the meaning of my existence?", "What happens after death?", "Do others see me as I see myself?", "Am I wasting my life?" These questions — all products of the cortical expansions of sapiens — do not improve evolutionary fitness. They do not increase reproductive success. They do not prolong species survival. They are cognitive noise generated by brains that grew too large.

Traditional Physical Anthropology interprets these changes as linear advance: a larger brain permits superior cognition, which translates into better adaptation. But this is anthropocentrism disguised as Science. If we measure evolutionary success by the criterion that matters — species longevity — erectus, with its 700-1,250 cm^3, was astonishingly more successful than sapiens with its 1,200-1,500 cm^3. The expanded anatomy of sapiens does not represent progress. It represents an evolutionary experiment whose viability remains unproven on any meaningful timescale.

The brain is not a cheap organ. That obvious statement takes on staggering proportions when we quantify the energetic cost of the expanded neuroanatomy of ***Homo sapiens***. The adult human brain, weighing approximately 1.4 kg — about 2% of total body mass — consumes between 20% and 25% of resting metabolic energy. This disproportion is extraordinary. No other organ, not even during intense activity, consumes such a large fraction of available resources. The heart at peak effort consumes about 10% of energy. Skeletal muscles at rest, about 20%. But the brain, even during deep sleep, maintains its absurdly high metabolic demand.

For *Homo erectus*, with a smaller brain and less expanded cortices, the cerebral metabolic cost was significantly lower — estimated at between 12% and 15% of total resting energy, based on the allometric scaling models developed by Aiello & Wheeler (1995) in their expensive tissue hypothesis, and refined by subsequent analyzes (Leonard *et al.*, 2003; Navarrete *et al.*, 2011) that confirmed the broad trend while complicating the precise trade-off between brain size and gut size. A difference of 5% to 10% may seem small, but in evolutionary terms it is enormous. It means sapiens must consume more calories simply to keep the brain functioning, regardless of physical activity. In environments where resources were limited — and throughout most of human evolutionary history, they were — that extra energy demand represents a severe adaptive disadvantage.

The cost is not only caloric. It is nutritional. The brain consumes glucose almost exclusively as fuel. It does not store energy as fat or glycogen in significant quantities. By contrast, other high-demand organs are metabolically flexible: the heart, for example, preferentially oxidizes fatty acids and can efficiently use ketone bodies such as acetoacetate and β-hydroxybutyrate, especially under fasting or high-demand conditions. The brain has no such flexibility. It depends on a constant supply of glucose arriving through the bloodstream, and any interruption — even brief — causes rapid cognitive dysfunction. Moderate hypoglycemia compromises judgement, attention, processing speed. Severe hypoglycemia causes confusion, seizures, coma. The sapiens brain, with its increased energy demand, is even more vulnerable to fluctuations in glucose supply.

Beyond glucose, the brain requires specific fatty acids — particularly long-chain omega-3 fatty acids like DHA — to maintain neuronal membrane integrity. The extensive myelination of the expanded cortices of sapiens demands significant quantities of specific lipids. Deficiency during development results in permanent cognitive deficits. Erectus, with less expanded cortices and lower myelination demands, had less exacting nutritional requirements. It could survive and reproduce successfully even in environments where lipid resources were relatively scarce.

The central evolutionary question is: what did sapiens gain in exchange for this disproportionate energetic investment? The usual answer — "superior cognitive capacity that allowed us to dominate the planet" — is tautological and anthropocentric. Dominating the planet is not a measure of evolutionary success if that domination results in accelerated extinction; we do not even know whether we have dominated the planet, or what bacteria would think of that claim. And "superior cognitive capacity" is a vague description that conceals important asymmetries. Superior in what? For what purpose? By which criterion?

If we compare cognitive capacities specifically relevant to survival in natural environments — long-range spatial navigation, resource-location memory, reading of environmental signals, cooperative hunting coordination, prolonged physical endurance — erectus was not cognitively inferior to sapiens. It may have been superior in several of these dimensions, as suggested by comparative ecological and archaeological evidence. Its attention, not fragmented by recursive metacognition, was more available for processing relevant environmental information, insofar as current evidence suggests it lacked the neural machinery for self-monitoring loops. Its spatial memory, not burdened by ruminations about a distant autobiographical past, was more efficient for immediate adaptive tasks. Its capacity for group coordination, not complicated by a hypertrophied theory of mind that increases the cognitive load of social inference, permitted more fluid cooperation.

What sapiens gained was a set of cognitive capacities that are superior only when measured by anthropocentric criteria: the ability to construct complex abstract narratives, to plan distant hypothetical futures, to create elaborate symbolic systems, to do science, philosophy, art — and, as a side effect, to devise intricate schemes against members of its own group in pursuit of status or power. A talent, admittedly, of limited adaptive value. These are impressive achievements. But their adaptive value becomes dubious when we weigh costs against benefits on an evolutionary scale. A million and a half years of erectus success *versus* three hundred thousand years of sapiens — with uncertain prospects of reaching five hundred thousand — suggest that the "superior cognitive capacities" of sapiens may not be superior at all when the criterion is species longevity.

There is a useful analogy here with maladaptive hypertrophy in other species. The most cited example is the irish elk (*Megaloceros giganteus*), but we will return to it later. For now, consider a simpler example: peacocks. The male peacock's extravagant tail consumes significant resources to grow and maintain, compromises mobility, and increases visibility to predators. It exists only because peahens prefer males with large tails — sexual selection in action. But if the environment changes rapidly and predation pressure increases, that tail, once a reproductive advantage, becomes a lethal disadvantage. The ornamental hypertrophy cannot reverse quickly enough, and the population collapses. The analogy is illustrative, not literal, but it captures the logic of costly traits under shifting conditions — conditions that are themselves contingent, shaped by geological, ecological, and cosmic accidents rather than any stable or predictable trajectory.

The sapiens brain can be understood as a similar hypertrophy. It expanded not necessarily because greater size conferred direct adaptive advantages, but possibly through complex sexual and social selection processes. Individuals with greater language capacity, more elaborated theory of mind, greater capacity for resource accumulation and social manipulation may have held reproductive advantages within increasingly complex social structures. But these advantages are context-dependent. If the social context destabilises, if resources become scarce, if the external environment imposes more direct pressures, the cognitive hypertrophy that seemed advantageous becomes dead weight.

And the cost is not only energetic or nutritional. It is obstetric. A large brain requires a large skull. A large skull requires a wide pelvis for passage during birth. But a wide pelvis compromises the efficiency of bipedal locomotion. Sapiens resolved this dilemma — commonly referred to as the "obstetric dilemma," though contemporary research suggests it involves multiple interacting constraints — through a ruthless solution: we are born neurologically premature. A foal is born and within hours is running. A chimpanzee infant is born with a brain already 40% of adult size. A human infant is born with a brain only 25% of adult size, completely defenseless, requiring intensive parental care for years.

This extended dependence is sometimes romanticized as "neoteny," or more precisely "heterochrony," permitting greater learning plasticity. But in evolutionary terms the vulnerability is enormous. The erectus infant was also relatively immature compared to other primates, but not to the extreme degree of sapiens, as far as fossil evidence suggests. The sapiens infant, whose brain will continue growing disproportionately after birth, is radically more vulnerable. Archaeological data show that infant mortality in pre-modern populations was astronomical — frequently above 30% before age five. This is the direct cost of cerebral hypertrophy: not only does it consume more energy throughout life, but it also compromises the viability of the earliest stages of development.

There is also the thermal cost. A working brain generates heat. A great deal of heat. The demand of 20-25% of metabolic energy means that 20-25% of body heat is generated in the skull. This requires efficient cooling mechanisms. Sapiens developed complex vascular systems to dissipate cerebral heat — emissary veins that allow blood to cool before reaching the brain, a complex carotid network functioning as a cooling system. But these mechanisms have limits. In very hot environments, during high fever, or during intense physical activity, the sapiens brain risks overheating. Erectus, with a smaller brain generating less metabolic heat, likely held an adaptive advantage in the hot tropical environments where the species evolved.

When we sum all these costs — energetic, nutritional, obstetric, thermal — and place them on the scale against the supposed adaptive benefits of expanded cognitive capacity, the evolutionary arithmetic does not balance as easily as the triumphalist narrative suggests. Sapiens staked an enormous quantity of evolutionary resources on an organ that consumes disproportionately, generates multiple vulnerabilities, and offers in return a capacity whose adaptive utility remains uncertain. This is not linear progression toward superior forms of consciousness. It is a risky evolutionary experiment whose outcome remains open — and untested on the timescale that matters: deep time.

Acheulean technology emerged roughly 1.7 million years ago. It consisted of a repertoire of techniques for producing bifacial hand-axes — stone tools flaked on both sides to create symmetrical cutting edges: the biface. Homo erectus mastered this technology and used it for more than a million years. More than a million years. Let that number resonate: the same technology, essentially unchanged in its fundamental principles, persisted across spans of time that make the entire history of sapiens civilization resemble the blink of an eye.

When nineteenth-century archaeologists began finding Acheulean axes on different continents, separated by hundreds of thousands of years, they initially suspected errors in dating. How could a technology remain stable for so long? How could a species manufacture essentially the same tool for a million years without "improving" it, without "evolving" it, without replacing it with something "more advanced"? The answer that prevailed for decades was simple and flattering to us: erectus was cognitively limited, incapable of innovation, trapped in rigid behavioral patterns that bordered on instinct.

That interpretation reveals more about our own biases than about erectus. We projected our compulsion toward innovation as a universal standard of progress, then interpreted the absence of that compulsion as cognitive deficit. But there is an alternative explanation that makes far more evolutionary sense: Acheulean technology persisted for a million years because it worked. Because it was the optimal solution to the problems erectus faced. Because it did not need to be improved.

A well-made Acheulean hand-axe is an object of functional elegance. Its bilateral symmetry is not ornament but the outcome of a knapping technique that maximizes cutting efficiency and durability. The proportions between length, width and thickness are not

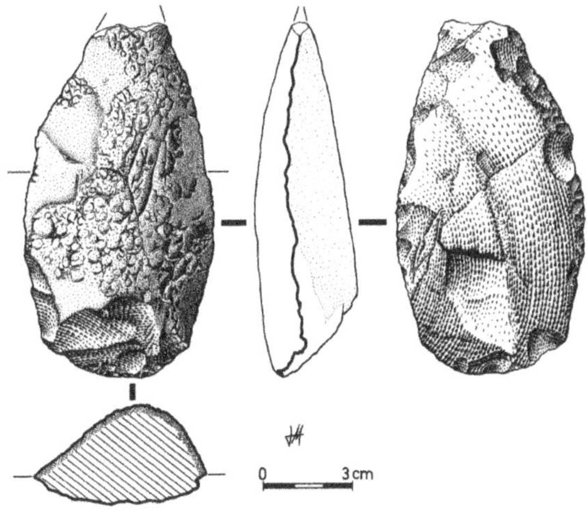

Acheulean bifacial hand-axe, shown in multiple views. The bilateral symmetry and proportions reflect functional optimization for cutting efficiency and durability. The design remained stable for over a million years, indicating adaptive robustness rather than cognitive stagnation.

arbitrary; they represent a balance between weight, force and cutting capacity refined over tens of thousands of years through cumulative trial and error. The choice of stone was not accidental: flint, obsidian, quartzite — each with specific fracture properties that erectus understood intimately and exploited with precision.

Technological stability does not indicate cognitive stagnation. It indicates the attainment of a plateau of adaptive excellence at which further change offers no significant benefit. Engineers understand this well: when a design reaches a local optimum in a space of possibilities, modifications tend to worsen performance rather than improve it. The Acheulean hand-axe was locally optimal and remarkably robust. It cut meat, broke bones, processed vegetables, worked wood, and served occasionally as a defensive implement. It was an exceptionally effective multi-purpose tool. Why replace it?

Contrast this with the technological trajectory of sapiens. From the emergence of the first technologies specifically associated with our species — the Upper Palaeolithic lithic industries, roughly 50,000 years ago — to the present, the rate of technological change has accelerated exponentially. Stone tools gave way to metal tools. Metal tools to simple machines. Simple machines to complex machines. Complex machines to computational machines. Computational machines to artificial intelligence systems. Each transition occurred on a progressively shorter timescale. The acceleration is so extreme that a sapiens born only a century ago, transported to the present, would be utterly disoriented by contemporary technology.

The standard narrative interprets this acceleration as triumph — proof of cognitive superiority, evidence of irreversible progress. But does nature favor progress? There is an equally valid, perhaps more precise interpretation: the frenetic innovation of sapiens is not a symptom of superiority but of inadequacy. It is an attempt to resolve through external means problems that are fundamentally internal. It is the pathological search for "more," a search that can never be satisfied because the problem is not technological but psychological.

A clinical parallel is instructive: chemical dependency. The dependent requires increasing doses of a substance to obtain the same effect. The brain adapts to the presence of the drug, developing tolerance. The initial pleasurable effect diminishes. The dose must be increased. But the increase produces ever more fleeting satisfaction, demanding still larger doses. It is an endless spiral that ends in overdose or collapse.

Sapiens behaves in a similar way with respect to technology. Each innovation promises to solve our problems once and for all. Agriculture would end the uncertainty of hunter-gatherer subsistence. Did it? It created civilizations dependent on monocultures vulnerable to collapse, generated brutal social hierarchies, and fueled wars of territorial conquest. Industrialization would end material scarcity. Did it? It produced labor alienation, planetary-scale pollution, and compulsive consumerism that never satisfies. The digital revolution would solve the problems of information and communication. Did it? It fragmented attention, fueled epidemics of anxiety and depression, and eroded the capacity for deep concentration.

With each cycle, the promise is the same: this time it will be different; this time technology will truly solve the problem. And with each cycle, new problems emerge — often worse than the originals — demanding still more technology. It is the existential hamster wheel spinning ever faster without moving from the spot.

Erectus did not share this compulsion. It made its Acheulean hand-axe, used it for years, perhaps decades, and eventually replaced it with another essentially identical one. Not because it lacked the cognitive capacity for innovation — the technical complexity of bifacial knapping is considerable, requiring sequential planning, fine motor coordination, and a three-dimensional understanding of fracture propagation in different stones. Erectus possessed the cognitive resources necessary for innovation within the domains relevant to its ecological niche. What it lacked was the neurotic restlessness that drives sapiens to seek "better," "newer," "more advanced."

That restlessness has clear neuroanatomical roots. The expanded prefrontal cortex of sapiens generates the capacity for elaborate anticipation of the future. But anticipating the future is a double-edged sword. It enables planning, but it also generates chronic anxiety. Sapiens lives in a permanent state of preparation for hypothetical futures — always imagining worse scenarios, always wanting to be more prepared, more secure, more protected. This anticipatory anxiety drives technological innovation: if we have better tools, we will be safer. But safety never arrives, because new threatening hypothetical futures continue to be generated indefinitely.

Erectus lived predominantly in an extended present. Not because it was incapable of anticipation — it certainly planned hunts, foresaw seasonal changes, and prepared for migrations — but because its anticipation was concrete, tied to tangible near-future events. It did not experience the kind of abstract, future-oriented rumination that keeps sapiens awake at night. It lacked the neurological machinery for that order of temporal projection. And paradoxically, this was an adaptive advantage. It allowed cognitive energy to be invested in responding effectively to the present environment rather than dissipated in anxieties about futures that might never materialize.

The capacity of sapiens to plan distant futures is frequently cited as a major adaptive advantage. But when examined empirically, that advantage becomes far less obvious. Studies of contemporary hunter-gatherer societies — imperfect but informative analogues for how erectus and pre-agricultural sapiens likely lived — show that they do not plan much beyond a few weeks. They do not accumulate significant surpluses. They do not build structures intended to last generations. They live in what anthropologists call "original affluence": working relatively little, four to six hours a day, enjoying diverse and nutritious diets, and knowing their environments intimately. The capacity for long-term planning becomes an "advantage" only in the context of sedentary agricultural societies — and these appeared merely 10,000 to 12,000 years ago, a recent moment in the history of sapiens.

For the first 280,000 years of sapiens existence, the expanded capacity for long-range future planning — projection across decades, not the medium-term anticipation already evident at Blombos Cave and Howiesons Poort — conferred no clear adaptive advantage over the more present-centered orientation of erectus. The archaeological record shows behavioral modernity emerging in bursts, not as a continuous ramp, which suggests that whatever cognitive capacities underlay it were not continuously under positive selection. What it conferred was a psychological cost: anxiety, worry, chronic dissatisfaction. And it is this chronic dissatisfaction that drives frenetic technological innovation — not because innovation improves evolutionary fitness, but because it offers temporary relief from internal restlessness.

A thought experiment clarifies the point. Imagine two groups facing the same natural environment: one composed of erectus with stable Acheulean technology, the other of sapiens in constant technological innovation. Which group survives longer? The answer is not obvious. The sapiens group may develop technologies that offer temporary advantages — better weapons, better shelters, better food preservation. But these advantages come with costs: greater resource consumption, greater dependence on supply chains vulnerable to disruption, greater fragility in the face of civilizational collapse when complex systems fail. The erectus group, with technology stable and tested over a million years, possesses robust resilience. It does not depend on complex supply chains and is not vulnerable to the collapse of interdependent systems. It can survive across a wide variety of environments using essentially the same toolkit. On a scale of thousands of years, that resilience may matter more than innovation. And empirically it did: erectus survived 1.5 million years. Sapiens has survived only 300,000, and its future prospects are increasingly uncertain.

The fundamental difference is not cognitive capacity for innovation but psychological disposition toward the present. Erectus appears to have operated within a behavioral equilibrium: its tools worked, its hunting strategies worked, its social organization worked. Sapiens is never satisfied — always wanting more, better, different. That dissatisfaction, romanticized as "the human spirit" or "curiosity," is a neurological symptom of cortical hypertrophy: a brain generating expectations that cannot be satisfied, creating an existential void that no technology can fill.

The *Megaloceros giganteus*, the irish elk, carried on its head the metaphor of its fate. The antlers of this Pleistocene cervid reached spans of up to 3.65 metres and weighed up to 40 kilograms. They were spectacular bony structures, covered in velvet during growth, branching into complex palmations that made each male recognizable at a distance. They were also weapons of limited practical use — too heavy for efficient maneuvering, vulnerable to fracture under impact, and voracious consumers of calcium and other minerals that had to be mobilized each year to produce a new set.

The evolutionary history of *Megaloceros* is instructive. The antlers expanded over hundreds of thousands of years through a combination of sexual selection and allometric scaling — their size was partly a consequence of the animal's massive body (Lister, 1994; Moen *et al.*, 1999), though the degree of hypertrophy exceeded what allometry alone would predict, pointing to a directional sexual-selection component superimposed on the scaling relationship. Females preferred males with larger antlers, likely because antler size honestly signaled genetic quality and resource-acquisition capacity: a male capable of growing enormous antlers was well nourished, strong, and genetically robust. Female preference created a directional selective pressure — a pressure that didn't merely nudge the trait but shoved it forward generation after generation. Males with larger antlers reproduced more, exported the underlying genetic architecture, and the population mean marched inexorably upward. It wasn't refinement; it was escalation. A runaway. The kind of evolutionary momentum that doesn't ask whether the trait still makes sense — it just keeps pushing because the system has no brakes.

But runaway selection has a fundamental problem: it can produce traits that enhance individual reproductive success while undermining

The Great Irish Elk (***Megaloceros giganteus***), illustrated by H. N. Hutchinson. Its antlers could reach spans of up to 12 feet, a dramatic example of runaway directional sexual selection driving morphological hypertrophy beyond ecological viability.

species survival. The enormous antlers of ***Megaloceros*** worked well in contexts of male competition and female choice during periods of resource abundance. But when the environment changed — the climate cooled, forests expanded, open grasslands shrank — those antlers became a lethal disadvantage. Navigating dense forest with 3.65 metres of antler is a biomechanical ordeal. Growing 40 kilograms of new bone each year requires quantities of calcium and phosphorus that may not be available when vegetation changes. Carrying such weight compromises balance and speed — critical for avoiding predators.

Megaloceros could not reverse the hypertrophy of its antlers quickly enough. Evolutionary processes do not "concern themselves" with reversing traits simply because they have become maladaptive; and even if they did, they would encounter resistance, because the genes producing large antlers are intertwined with other physiological characteristics. It is not possible to "switch off" antler growth without affecting other traits. Moreover, as long as females continued to prefer males with large antlers, sexual selection pushed in the opposite direction of natural selection. The result was an evolutionary trap: the antlers were harmful, but the species could not rid itself of them. ***Megaloceros*** went extinct approximately 7,700 years ago.

The analogy with the brain of ***Homo sapiens*** is obvious, though rarely appreciated in its full implications. Our expanded cortex, like the antlers of ***Megaloceros***, likely evolved substantially through sexual and social selection, not merely through direct environmental pressures. Within complex social groups, individuals with greater language capacity, more elaborated theory of mind, and greater skill in social manuvering held reproductive advantages. They were more persuasive, formed more effective alliances, and navigated social hierarchies more successfully. This reproductive advantage created selective pressure for an even larger cortex. A positive feedback loop may have emerged: larger brains created more complex social structures, which in turn selected for still larger brains capable of navigating those structures.

But, as with the elk's antlers, the hypertrophied brain of sapiens carries costs that may not be offset by real adaptive benefits. The metabolic cost is substantial, but additional costs become apparent when we shift perspective from intraspecific competition — individual against individual — to the viability of the species as a whole.

First, there is the ecological cost. Sapiens, with its large brain and expanded cognitive capacity, developed an unprecedented ability to modify environments. This capacity appears advantageous — it allows more efficient exploitation of resources, enabling us to adapt environments to our needs rather than adapt ourselves to the environment. But on longer timescales, this capacity is proving catastrophic. We have modified environments so radically that we are destroying the ecological systems on which we depend. Anthropogenic climate change, accelerating biodiversity loss, soil depletion, ocean pollution — all are direct products of the cognitive capacity that supposedly makes us superior.

Erectus, with more limited cognitive capacities, lacked the ability to modify environments on a large scale. This appears to be a disadvantage. But it meant that erectus remained integrated within ecological systems rather than destroying them. It lived within the limits imposed by natural regeneration. It could not fish entire populations to exhaustion. It could not fell forests faster than they regrew. It could not pollute rivers beyond their capacity for self-purification. These limitations, which we interpret as inferiority, were in fact guarantees of long-term sustainability.

Second, there is the psychological cost — already mentioned, but deserving deeper examination. The same cognitive capacities that allow sapiens to produce science, philosophy and art — all genuine achievements — also generate forms of psychological suffering that did not exist in erectus. Chronic anxiety, depression, dissociation, personality disorders, various addictions: all specifically human, all products of complex self-representation, elaborate temporal projection, and recursive metacognition.

Everything we know indicates that erectus experienced pain, fear and sadness in response to concrete events. But all available evidence also suggests that it lacked the neuroanatomical capacity for existential anxiety — that diffuse dread about the meaning of life not tied to any immediate threat. It lacked the capacity for clinical depression in the modern sense — that persistent emptiness that endures even when external conditions are favorable. It lacked the capacity for pathological dissociation — the distancing from oneself that emerges when metacognition becomes excessive. These forms of suffering are not cultural artifacts; they are direct consequences of the expanded neuroanatomy of sapiens. Cultures can modulate their expression, but cannot eliminate them, because they are rooted in cerebral structure. A sapiens, no matter how well adjusted or materially secure, remains vulnerable to these specifically human forms of suffering because they arise from the very type of consciousness our hypertrophied brain generates.

Third, there is the cost of cognitive vulnerability to manipulation. Large brains with complex symbolic capacities are structurally prone to having certain abstractions take hold — not because the organism "seeks" them, but because symbolic systems can latch onto narratives that spread through cultural contingencies, stochastic drift, or even sexual-selection dynamics. These capacities make sapiens susceptible to destructive ideologies, fanaticisms, and collective illusions with no correspondence to reality. Erectus could not be mobilized to die for abstractions such as "nation," "god," "race," or "progress." It lacked the symbolic-representational machinery required for such commitments. Its loyalties were concrete — to its immediate group, to individuals with whom it had direct bonds. Sapiens, by contrast, can be mobilized for genocidal wars in the name of abstractions. It can sacrifice concrete well-being for abstract promises. It can believe in narratives that contradict direct experience because symbolic thinking allows abstractions to substitute for reality.

This vulnerability is not a malfunction; it is an inevitable consequence of having a brain capable of complex representations detached from immediate experience. And it carries enormous evolutionary costs. Wars, genocides, religious fanaticisms, totalitarian ideologies — all specifically sapiens phenomena, all responsible for massive mortality throughout history. Erectus undoubtedly had conflicts, but it lacked the capacity for large-scale organized violence motivated by abstract ideology.

When we sum these costs — metabolic, ecological, psychological, cognitive vulnerability — and weigh them against real adaptive benefits (not benefits measured by anthropocentric criteria, but by an evolutionary criterion: species longevity), the account does not balance in favor of sapiens. The hypertrophied brain may be leading us toward the same fate as *Megaloceros*: extinction caused by a trait that appeared advantageous in a specific context but proved maladaptive when that context changed.

There is, however, one essential difference. *Megaloceros* lacked the capacity to recognize its own hypertrophy as a problem. It could not choose to have smaller antlers. It was trapped in an evolutionary dead end. Sapiens, with its metacognition and capacity for reflection, could theoretically recognize that the large brain is not an unmixed blessing — that our expanded cognitive capacities come with costs that may be prohibitive. We could, in principle, develop cultures that minimize the damage of cognitive hypertrophy: contemplative practices that reduce excessive rumination, social structures that limit destructive competition, economic systems that respect ecological limits.

And here lies a radical irony: the very cognitive characteristics that would allow us to recognize the problem are those that prevent us from solving it. The metacognition that permits understanding that a large brain is a problem also feeds narratives of human exceptionalism and anthropocentrism. The capacity for future planning that could orient us toward sustainability also generates the anxiety that drives compulsive consumption. The capacity for abstract thought that could formulate solutions also makes us vulnerable to ideological illusions promising salvation through "more" — more technology, more growth, more control.

Erectus was, in a certain sense, protected from itself by its cognitive limitations. It could not imagine grand utopian futures that would never materialize. It could not construct economic systems based on infinite exponential growth. It could not believe it was exempt from the laws of nature. It lived within limits because it lacked the cognitive capacity to imagine existing beyond them. We possess that capacity — and it may be precisely what condemns us.

We end up saying that more is less. And this is not rhetorical provocation. It is a precise description of actual evolutionary dynamics. More prefrontal cortex did not yield greater evolutionary fitness. More metacognitive capacity did not yield greater well-being. More technological innovation did not yield greater sustainability. In each case, "more" generated costs that may exceed benefits. And if history is any guide, species that enter regimes of "more is less" rarely survive long enough to reverse the process.

Erectus had less — less cortex, less metacognition, less technology — and survived a million and a half years. We have more. And will we last three hundred thousand? Five hundred thousand? A million? The evolutionary story remains open. But the data so far suggest that the irish elk is not merely a metaphor. It is a warning — a warning rooted in a transformation more fundamental than the volumetric expansion of the brain. Cortical hypertrophy made possible something qualitatively new: the recursive language that exiled us from the world erectus inhabited naturally. These limitations, which our anthropocentric perspective frames as inferiority, were in fact guarantees of long-term sustainability.

III.
The spell of words

The fact is that men refused to be what the past, in the likeness of animals, had proposed them to become. They made themselves inventors of worlds. And they planted gardens, raised huts, houses and palaces, built drums, flutes and harps, composed poems, transformed their own bodies by covering them in pigments, metals, marks and cloth, invented banners, erected altars, buried their dead and prepared them for the journey — and in their absence, sang laments through the days and through the nights.

RUBEM ALVES

The erectus looked at the world and saw what was there: stone, river, prey, sky. The sapiens looked at the same world and saw infinite possibilities, each thing ready to be transformed into something else through the word. The stone became an altar, the river became a deity, the prey became a totem, the sky became a promise. Rubem Alves celebrates this capacity as a poetic achievement of humanity. But every form of magic has a price: when anything can become something else through language, nothing remains simply what it is. We lost the ability to dwell in the world as it presents itself. We gained the spell of words, and found ourselves confined within meanings of our own making, trapped not by the real but by the symbols that stand between us and it.

There is a terminological confusion that has accompanied the philosophy of language for centuries, and it must be dissolved before we proceed. To avoid ambiguity, I adopt a four-part distinction. I use "language" to refer to the biological faculty of signaling and communication present across living systems — from bacteria to bees to primates. I use "a language" to refer to a specific human idiom such as english, yoruba, or japanese. Following Ferdinand de Saussure, I retain the french term *langue* to designate the abstract, collective symbolic system shared by a community. And I use *parole* to refer to the individual act of speech, the concrete realization of that system in a particular moment. These four terms are not interchangeable. They describe different layers of a phenomenon that English unfortunately collapses under a single word: "language."

Language, in this broad biological sense, permeates life. It is not a human invention, nor even an animal invention. It is a fundamental characteristic of complex organisms that must coordinate behavior through signals.

Bacteria practice *quorum sensing* — an extraordinarily sophisticated form of chemical communication through which populations of unicellular organisms detect their own density. When the concentration of specific signaling molecules reaches a critical threshold, genes responsible for collective behaviors are activated. Bacteria "know" when there is sufficient critical mass to form biofilms, to express virulence factors, to initiate coordinated bioluminescence. This is language in the functional biological sense: a system of signals carrying information and modifying the behavior of receivers. No consciousness is involved, no intention in the human sense, but the communication is real and operational.

Bees practice three-dimensional spatial communication of remarkable precision. The round dance performed by a bee returning to the hive after finding a food source is not random behavior. It is a code transmitting the distance, direction, and quality of the resource. The angle of the dance relative to the vertical corresponds to the angle between the sun and the food source. The duration and intensity of the movement transmit distance. The vigor of the dance signals the richness of the resource. Other bees decode this information and fly directly to the indicated location, adjusting the route as the solar position changes. This is language: a conventional system of signals enabling the transfer of complex information.

Non-human primates possess referential vocalizations — specific calls for different types of predators. Vervet monkeys, studied extensively by Dorothy Cheney and Robert Seyfarth, emit distinct alarms for eagles, leopards, and snakes. Each alarm provokes an appropriate behavioral response: looking up and running to lower ground upon hearing the eagle alarm, climbing a tree at the leopard alarm, standing erect and scanning the ground at the snake alarm. These calls are not mere emotional expressions: they are signals referring to specific external objects and provoking coordinated reactions in the group. This is language: a system of signals with shared meaning that modifies collective behavior.

Homo erectus had language in this sense. As far as current evidence allows us to infer, it possessed vocal communication coordinated with gesture, facial expression, and body posture. It had conventional signals for alerting to dangers, coordinating hunts, transmitting stone-knapping techniques, organizing the division of tasks within the group. It likely had a proto-language — a communication system more sophisticated than the referential vocalizations of other primates but less elaborate than the fully recursive langue of sapiens. It could combine signals in limited ways, refer to temporarily absent objects, and teach techniques by demonstration accompanied by vocalization. All of this constitutes functional language adequate for complex social organization and cultural transmission.

What sapiens developed is categorically distinct. It is not merely a more complex or sophisticated language along a quantitative continuum. It is a qualitative leap into something fundamentally different: a *langue* — an arbitrary, recursive, generative symbolic system with unlimited capacity for temporal and spatial displacement. The difference between the language of erectus and the *langue* of sapiens is not like the difference between a simple calculator and a powerful computer. It is like the difference between a calculator and a machine capable of simulating any possible calculator, including itself simulating other calculators. It is the difference between a tool and a meta-tool that operates on tools, including on itself.

Recursivity is the defining characteristic of the human *langue* and its most consequential property. Recursivity means the capacity to embed structures within structures of the same type, indefinitely. In syntax, it means embedding clauses within clauses: "The man who saw the dog that bit the cat that chased the rat that ate the cheese that was in the house that Jack built died yesterday." Each relative clause is embedded within another, creating a hierarchy of dependencies requiring intensive cognitive processing simply to decode. But syntactic recursivity is only the surface manifestation of a deeper conceptual recursivity.

We can have thoughts about thoughts. We can think about the fact that we are thinking. We can ask ourselves why we are asking ourselves something. We can have intentions about our own intentions. We can monitor our own monitoring. This capacity for representing our own representations, full meta-representation, depends on the recursive structure of human *langue* and on the expanded prefrontal-parietal network characteristic of ***Homo sapiens***. Nothing in the neuroanatomy of erectus, as revealed by endocasts, suggests the presence of the cortical volume or connectivity required for such recursive metacognition. Erectus could think "there is an antelope there" and act appropriately. Sapiens can think "I am thinking about the possibility that there is an antelope there, but I may be mistaken, so I should examine my reasons before acting." The difference is not trivial. It is an abyss between worlds.

Noam Chomsky argued that recursivity is the central property distinguishing human *langue* from all other animal communication systems — a claim contested by Daniel Everett (2005), who described the amazonian Pirahã language as operating without recursive clause embedding. Whether Pirahã truly lacks recursion or merely deploys it differently remains debated, but the controversy need not concern us here: what matters for the present argument is not whether every single human langue is recursive, but that the overwhelming majority are, and that the cognitive architecture of sapiens supports recursion as a default mode of symbolic processing.. The capacity to generate discrete infinity from finite means, producing an unlimited number of new sentences from a finite vocabulary and finite grammatical rules, depends fundamentally on syntactic recursivity. Once you have the capacity to embed clauses within clauses, you can continue embedding indefinitely, limited only by constraints of memory and attention, not by structural limits of the system itself.

But Chomsky, focused on formal syntax, did not fully explore the psychological consequences of this capacity. Syntactic recursivity permits conceptual recursivity. And conceptual recursivity is a double-edged weapon. It permits science — the capacity to construct theories about theories, meta-analyses, models that model other models. It permits philosophy — reflection on reflection itself, epistemology that questions the foundations of knowledge. But it also permits pathological rumination — worry about worries, anxiety about anxiety, fear of one's own fear, guilt about feeling guilty. Erectus, as far as current evidence indicates, lacked the recursive *langue* and the prefrontal–parietal architecture required for self-referential psychological suffering. In that sense, it was protected from these forms of torment that depend on meta-representation.

The arbitrariness of the linguistic sign is the second distinctive characteristic of the human *langue*. Ferdinand de Saussure demonstrated that the relationship between the signifier (sound or visual sequence) and the signified (concept) is predominantly arbitrary, established by social convention rather than natural necessity. There is nothing in the sound sequence 'tree' that resembles actual trees. Partial exceptions exist, onomatopoeia, sound symbolism, the *bouba*/kiki effect documented by Ramachandran and Hubbard (2001), but these are marginal phenomena that confirm the rule: the overwhelming majority of signs in any human langue are conventional, not iconic. Different languages use completely distinct signifiers for the same concept: tree, arbre, baum, árbol, albero. If the relationship were natural, all languages would use the same sound. They do not, because the relationship is arbitrary.

This contrasts fundamentally with non-linguistic communication. The alarm vocalizations of primates are not entirely arbitrary — they have acoustic properties that relate functionally to context. The alarm for an aerial predator tends to be short and sharp, facilitating rapid localization of a threat from above. The alarm for a terrestrial predator tends to be lower and more prolonged. There is partial iconicity: the signal's characteristics reflect the situation's characteristics. But in the human *langue*, that iconicity is minimal. Words are pure symbols, arbitrary conventions that function only because the community of speakers tacitly agrees that a given sound means a given thing.

This arbitrariness has a profound consequence: since symbolic signs are not anchored to reality by natural resemblance, they can drift indefinitely from concrete experience. We can have words for things we have never seen, places we have never visited, events that never happened. We can have concepts like "infinity," "eternity," "absolute nothing" — concepts corresponding to no possible experience, yet perfectly manipulable linguistically. We can construct narratives completely detached from any factual reality and inhabit those narratives as though they were parallel worlds. Erectus, with a more iconic and less arbitrary communication system, was anchored in direct experience in ways that sapiens no longer is.

Temporal and spatial displacement is the third distinctive characteristic. Human *langue* allows reference to objects, events, and situations remote in time and space. We can speak about what happened a thousand years ago. We can speak about what will happen a thousand years from now. We can speak about places on the other side of the planet. We can speak about completely imaginary places that exist nowhere. Animal language has limited displacement — bees can communicate the location of food several kilometers away through the waggle dance, while primates can produce alarm vocalizations referring to a predator that was present seconds ago. But human *langue* has unlimited displacement — we can speak about any place, any time, any possibility, real or imaginary.

Vygotsky understood that the linguistic system, what Saussure would call *langue*, is not merely a vehicle for expressing pre-existing thoughts. It restructures cognition. Children who have not yet mastered a language think differently from children who have. Pre-linguistic thought is situated, concrete, bound to immediate perception. Linguistic thought is decontextualized, abstract, capable of operating on symbolic representations independently of the sensory stimuli present. Vygotsky called this "verbal thought" — a qualitatively distinct form of cognition that emerges when *langue* is internalized and becomes the structure of thought itself.

Wittgenstein, in the *Tractatus* (5.6), stated: "The limits of my language mean the limits of my world." This does not mean that experience beyond language does not exist — Wittgenstein himself acknowledged, in the closing propositions, that there is something whereof one cannot speak. What it means is that we cannot articulate pre-linguistic experience as definite propositional thoughts, deliberately manipulate it, or communicate it to others. *Langue* does not merely express reality — it constitutes discursive psychological reality: the reality we can think about, reason with, and transmit through symbols. *Langue* does not merely express reality — it constitutes psychological reality. The phenomenological world of sapiens, mediated by recursive and symbolic *langue*, is radically different from the phenomenological world of erectus, which lived predominantly in direct experience not linguistically mediated.

And here is the central point that dissolves any romanticization of human linguistic capacity: *langue* does not only expand cognitive possibilities; it also creates cognitive vulnerabilities nonexistent in organisms without recursive symbolic systems. The same capacity that permits science permits pseudoscience. The same capacity that permits philosophy permits sophistry. The same capacity that permits abstract empathy with the suffering of distant strangers also permits the construction of abstract categories of "enemies" that justify genocide. *Langue* is a neutral tool in the same sense that a knife is a neutral tool — it can cut food or cut a throat, depending on who wields it and with what intention.

To name is to create. When we give something a name, we are not merely labeling a pre-existing reality. We are carving the continuous flow of experience into discrete units, imposing a categorical structure on reality that does not, in itself, come pre-divided into categories. Nature does not contain "trees" as a natural category. It contains photosynthetic woody organisms with immense morphological diversity. We created the category "tree," grouping some of these organisms and excluding others based on arbitrary criteria — height, trunk thickness, branching. Large shrubs are left out. Palms are included or excluded depending on how we define "tree." The category is a linguistic construction, not the discovery of a natural essence.

This becomes far more consequential when applied to psychological and social concepts. Consider "death." There is a biological event, the cessation of metabolic processes, the decomposition of tissue, the end of neural activity. Erectus witnessed this event. It saw companions die, saw animal carcasses, and understood death as a concrete present phenomenon. But it did not have the abstract concept of "death" detached from specific deaths. It could not think about "death" as a universal category, as the inevitable destiny of every living being, as an existential horizon that permanently defines the meaning of life. It could not fear death in general — only fear concrete present threats that might cause a specific death at that moment.

Sapiens, with a langue permitting temporal displacement and conceptual recursivity, created the abstract concept of "death" and began to inhabit it psychologically. We can think about our own future death, imagine what dying will be like, anticipate the absence of experience, feel existential terror about the inevitable annihilation that has not yet happened and may not happen for decades. This terror is not an adaptive response to a concrete present threat. It is a linguistic product — a symbolic construction that exists only because we have the word "death" and the recursive capacity to apply it to ourselves as an object of contemplation. We created a permanent object of terror by naming something that, without abstract naming, would be merely a series of occasional concrete events.

Consider the "self." There is a physiological continuity — a body that persists through time, memories that accumulate, behavioral patterns exhibiting relative stability. But there is no substantial entity "self" that remains identical through transformations. Every cell in the body is replaced multiple times over decades. Neurons die and synaptic connections constantly reorganize. Memories are rewritten each time they are accessed. Personality changes dramatically between childhood, adolescence, adulthood, and old age. The "self" of now is not the same "self" of ten years ago in any substantial sense beyond narrative convenience.

But langue creates the illusion of permanent identity by supplying the pronoun "I," which we consistently use to refer to ourselves across time. We construct an autobiographical narrative — a story about "me" as a continuous character traversing events, possessing stable characteristics, pursuing long-term goals, maintaining an essential identity. This narrative is a useful linguistic fiction for social navigation, but it is fiction. Erectus had physiological continuity, had memories, had habitual behaviors. But it did not have an elaborately linguistically constructed autobiographical narrative about the "self" as a permanent entity. It did not suffer from identity anxiety — that distinctively human concern about "who I really am," about authenticity, about coherence between past and present, about the long-term purpose that gives narrative meaning.

Consider "others" as an abstract category. There are concrete individuals — specific people with faces, voices, particular behaviors. Erectus interacted with individuals within its group, recognized them individually, and had differentiated relationships with each. It could experience conflicts, alliances, cooperation, all based on concrete interactions. But it lacked the capacity to construct the abstract category "others," defined by arbitrary symbolic characteristics and applicable to millions of individuals it had never personally encountered.

Sapiens creates categories like "jews," "tutsis," "communists," "infidels," "barbarians" — linguistic labels grouping diverse individuals under an abstract concept that allows treating them collectively, not as specific people but as instances of a category. Once the category is linguistically constructed and associated with abstract negative attributes, it becomes possible to hate, fear, or wish to eliminate "the others" without ever having concretely interacted with any individual belonging to the category. Genocide becomes possible only because langue permits creating an abstraction that separates the category from concrete individuals. You cannot kill millions of people you know personally — concrete empathy prevents it. But you can kill millions when they are reduced to an abstract category, linguistically constituted as a threat.

Racism, sexism, xenophobia — all depend fundamentally on *langue*. Not because langue "causes" prejudice in a simplistic sense, but because it provides the conceptual tools necessary to transform superficial observable differences into imaginary essences that justify differential treatment. Richard Lewontin demonstrated that, for individual genetic *loci*, variation within populations exceeds variation between populations — a finding that subsequent analyses have complicated (Edwards, 2003) by showing that multivariate clustering can approximate geographic ancestry. But even that clustering does not recover the discrete, essentialized categories that the word "race" implies. The linguistic category precedes and exceeds the biological data: it is not a description of nature but a narrative imposed on it. But "race" is a supremely powerful linguistic category that, once constructed and naturalized through repeated discourse, permits organizing perception, justifying hierarchies, and mobilizing collective violence. Erectus likely exhibited in-group preference — a widespread pattern among social primates. But it lacked the symbolic machinery required to construct racial ideologies based on abstract categories and to transmit them culturally across generations through a language.

Consider abstract moral concepts like "sin," "virtue," "duty," "justice." There are no concrete events corresponding to these words.

There are specific behaviors we approve or disapprove of. There are consequences of actions that benefit or harm others. But there is no substance "sin" floating in the world, no objective property "virtue" inherent in actions. These are linguistic concepts we created to categorize and evaluate behavior. Erectus had approval and disapproval of concrete behaviors within its group. It had implicit social norms maintained by reciprocity and punishment. But it had no abstract moral concepts detached from specific contexts.

Sapiens constructs elaborate moral systems — law codes, divine commandments, categorical imperatives, utilitarian calculations. These systems are all linguistic products. They permit moral judgment of actions remote in time and space, permit feelings of guilt about transgressions of abstract rules even when no concrete individual has been harmed, permit collective mobilization to punish violations of symbolic norms. They also permit forms of psychological suffering nonexistent without langue: abstract guilt, shame about not corresponding to linguistically constructed ideals, anguish about the incoherence between declared values and actual actions.

Langue does not merely name pre-existing psychological realities. It creates psychological realities that could not exist without the naming. Naming "anxiety" as a discrete internal state transforms a diffuse experience of discomfort into an identifiable psychological object about which we can have meta-anxiety — anxiety about being anxious. Naming "depression" as a medical condition transforms the experience of prolonged sadness into an identity: "I am depressed" rather than "I am sad right now." Naming is not neutral. It restructures experience. Erectus felt discomfort, fear, sadness. But it had no elaborate psychological vocabulary transforming emotions into objects of recursive contemplation.

All distinctively human suffering is linguistically mediated. This assertion does not deny that humans suffer physical pain, hunger, cold, injury — all sufferings we share with other animals and which erectus certainly experienced. But there are specifically human forms of suffering, impossible without recursive *langue*, that constitute the greater part of psychological torment in the modern era: existential anxiety, clinical depression, abstract guilt, identity shame, ideological hatred, comparative envy, fear of abstract death, and anguish about the meaning of life.

Existential anxiety is not fear of a concrete present threat. It is a state of diffuse apprehension about the meaning and purpose of existence, about the inevitability of death, about the possibility that life has no intrinsic meaning. It is a direct product of the linguistic capacity to formulate abstract questions like "Why do I exist?", "What is the meaning of everything?", "What happens after death?" These questions are linguistically constituted — they do not exist as psychological problems without a conceptual vocabulary that formulates them. Erectus did not have existential anxiety not because it had satisfying answers to these questions, but because it lacked the linguistic capacity to formulate them as questions.

Clinical depression, in the modern sense, is not merely intense sadness or prolonged grief. It is a state of anhedonia — the inability to feel pleasure — combined with negative rumination, a pessimistic view of the future, and feelings of worthlessness, frequently accompanied by suicidal ideation. Rumination is a central component: recursive thoughts about one's own inadequacy, about past failures, about the futility of future efforts. This ruminative recursivity is linguistically mediated. Severe depression frequently involves an incessant negative internal dialogue — a voice in the head repeating self-deprecating narratives. Without an internalized *langue* permitting internal dialogue, this form of rumination would not be possible.

Abstract guilt is not remorse for having harmed a specific person. It is the feeling of transgression in relation to abstract moral norms, frequently in situations where no concrete person has been harmed. Religious guilt about "impure thoughts," guilt about not corresponding to cultural ideals of success, guilt about structural privileges over which one has no individual control — all of these are forms of guilt mediated by abstract linguistic concepts. Erectus could feel remorse after harming a group member — an emotion observable in many social mammals. But it lacked the capacity to construct abstract moral codes and feel guilty for violating symbolic rules.

These assertions about the cognitive limitations of erectus are not arbitrary speculation. They are based on neuroanatomical evidence derived from endocasts — molds of the interior of the cranial cavity revealing cerebral morphology. Erectus endocasts show that the frontal and temporal lobes — the broad regions within which Broca's and Wernicke's areas are located in sapiens — were volumetrically smaller and morphologically distinct, with less pronounced convolutions in areas corresponding to the dorsolateral prefrontal and posterior parietal cortices (Holloway *et al.*, 2004; Falk, 2014). Endocasts do not resolve individual cytoarchitectonic areas, but the overall pattern of reduced frontal and parietal volume is consistent with limited capacity for recursive langue, elaborate metacognition, and the construction of autobiographical narratives. Without the neuroanatomical substrate for recursive *langue*, there is no way to internalize abstract moral codes transmitted linguistically. Without a prefrontal cortex with the volume and connectivity of sapiens, there is no way to maintain symbolic representations of norms detached from concrete contexts. The absence of abstract guilt in erectus is not a vague psychological inference — it is a direct consequence of documented anatomical limitations.

Identity shame is not the momentary embarrassment of a public error. It is the feeling of inadequacy in relation to linguistically constructed identity categories — shame about not corresponding to norms of masculinity, femininity, professional success, or physical attractiveness according to internalized cultural standards. These norms are transmitted linguistically, internalized through cultural narratives, and applied recursively through constant self-monitoring. Without a *langue* permitting the construction and internalization of abstract ideals, this form of shame would not exist.

Ideological hatred is not anger at an individual who has concretely harmed us. It is hostility directed at abstract categories of people, based on beliefs about essential shared attributes. Racial hatred, religious hatred, political hatred, all depend on a langue permitting the construction and dissemination of dehumanizing narratives about abstract "others." Erectus could feel anger at an individual who attacked it, or at a rival group competing for resources. But it lacked the capacity to hate abstractly millions of people it had never met, based on symbolic categories.

Comparative envy is not merely the desire to possess resources that another has. It is suffering caused by abstract social comparison — the feeling of inadequacy at perceiving that others possess more status, success, or recognition according to culturally constructed metrics. This comparison is linguistically mediated, narratives about "success," "failure," "falling behind in life," "wasting one's potential." Without a *langue* permitting the construction of autobiographical narratives and their comparison with idealized narratives, this form of suffering would be impossible.

The abstract fear of death, already discussed, is the existential terror of future annihilation linguistically formulated as an inevitable horizon. It is radically different from the adaptive fear of concrete present threats. Erectus felt fear when confronting a predator, a precipice, or fire, but it did not carry a permanent existential terror about the fact that it would eventually die anyway. That terror is a linguistic product — the concept "death" applied recursively to the narrative "self" as a future destination.

Anguish about the meaning of life is perhaps the most quintessentially human form of suffering. It is concern about whether life has purpose, whether we are living meaningfully, whether anything will remain after death. These concerns are entirely linguistically constituted; they correspond to no concrete adaptive problems, do not improve evolutionary fitness, and do not solve practical challenges. They are cognitive luxuries, or rather, cognitive costs, of possessing a *langue* capable of formulating abstract metaphysical questions.

Erectus did not suffer existential anguish about the meaning of life. Not because it had comforting answers provided by religion or philosophy — it had neither. It did not suffer this anguish because it lacked the linguistic capacity to transform the pragmatic question "What should I do now?" into the metaphysical question "What is the ultimate meaning of everything I do?" It lived immersed in the present, responding to concrete challenges, without a meta-narrative about the global purpose of existence.

A clear pattern emerges: distinctively human suffering arises from distancing ourselves from immediate experience through linguistic representation. Existential anxiety concerns abstract concepts rather than present threats. Depression involves rumination about past and future rather than responses to present circumstances. Abstract guilt concerns symbolic transgressions rather than concrete harms. Identity shame concerns failure to match linguistically constructed ideals rather than functional inadequacy. Ideological hatred targets abstract categories rather than individuals who pose real danger. Comparative envy concerns one's position in symbolic hierarchies rather than actual resource scarcity. The abstract fear of death concerns a future concept rather than a present threat. Anguish about meaning concerns metaphysical questions rather than practical problems. In every case, *langue* permits, and compels, us to leave direct experience and inhabit symbolic representations.

It is these symbolic representations, not concrete circumstances, that generate the greater part of human psychological suffering. People living in material comfort suffer profound depression. People in complete physical safety suffer paralyzing anxiety. People who have never been harmed by members of another group hate that group abstractly. People with health, resources, and relationships suffer anguish about the lack of meaning. This suffering is real; it is not illusory or trivial. But it is a linguistic product. It would not exist without the capacity to construct and inhabit symbolic representations detached from immediate experience.

Erectus was protected from these forms of suffering not by superior wisdom or better philosophical answers, but by cognitive limitations that prevented the formulation of the problems. Without recursive *langue*, there is no way to have thoughts about thoughts that generate rumination. Without abstract categories, there is no way to construct ideologies justifying hatred of millions of strangers. Without elaborate autobiographical narratives, there is no way to experience identity anxiety about the coherence of the "self" across time. Without abstract moral concepts, there is no way to feel guilt about symbolic violations. Without unlimited temporal displacement, there is no way to experience existential terror about abstract future death.

Langue, which we celebrate as humanity's supreme achievement, is also the instrument of psychological torture we carry internally and apply to ourselves incessantly. The spell we cast has imprisoned us. And there is no exit: we cannot voluntarily "switch off" *langue*, cannot return to pre-linguistic direct experience. Once *langue* is internalized and cognition structured by it, there is no way to think without it. Even the attempt to "live in the present" or "silence the mind" is linguistically formulated, mediated by verbal instructions, monitored by linguistic self-observation asking "Am I managing not to think?" — a question that is itself linguistic thought.

We are permanently exiled from the present by the *langue* that allows us to name the present as an abstract concept rather than inhabit it directly. Erectus lived in the only place where life actually happens: here, now, in this unrepeatable instant. We live in narratives about a past that no longer exists and projections about a future that does not yet exist, crossing the present as a narrow corridor between shadows and mirages, always on the way to some symbolic place we never reach because it exists only linguistically.

Rubem Alves celebrated the human capacity to transform stone into altar, river into divinity, body into poem. But he did not ask the price. The price is never being able to see stone simply as stone, river simply as river, body simply as body. The price is permanently inhabiting a second-order symbolic world where everything means something beyond itself, where nothing is sufficient in its raw presence, where interpretation always mediates experience. The price is the permanent exile from reality in favor of representations of reality. And in that linguistic exile, we suffer forms of torment that erectus, in all its "primitiveness," never knew.

AGORA

IV.
The lost present

> *The meaning of life: there is no question asked with greater anguish, and it seems that everyone is haunted by it from time to time. Is it worth living? The gravity of the question reveals itself in the gravity of the answer. Because it is not rare to see people plunged into the abysses of madness, or voluntarily choosing the abyss of suicide, for having obtained a negative answer. Other people, as Camus observed, let themselves be killed by ideas or illusions that give them reasons to live: good reasons to live are also good reasons to die.*
>
> **RUBEM ALVES**

Homo erectus did not ask about the meaning of life because it lived inside life, not outside it, observing it from a distance. The anguish described by Rubem Alves — that haunting question capable of driving someone to madness or suicide — is a privilege exclusive to a species that developed recursive language, the cognitive machinery that allows existence itself to become an abstract object of inquiry. Erectus had concrete reasons to live: hunger to satisfy, dangers to avoid, a group to protect, techniques to master. It did not need abstract justifications for life itself. We, sapiens, stepped out of the present — the only place where life actually happens — and began to inhabit it linguistically and temporally in every direction, turning existence into a philosophical problem that demands a solution. And the more we search for meaning, the more it slips away — because meaning appears only when we stop looking for it and simply live. But that is precisely the capacity we lost when we gained language.

Saint Augustine of Hippo, in *Book XI* of the *Confessions* (397-400), formulates one of the most penetrating analyses of the nature of time in western philosophy. He asks: "What, then, is time?" And answers with devastating honesty: "If no one asks me, I know; if I want to explain it to one who asks, I do not know." This is not rhetorical modesty. It is the recognition that time, when we attempt to grasp it conceptually, slips between the fingers of language. We live in time, but we cannot capture it in words without distorting it.

Augustine identifies a fundamental paradox: only the present actually exists. The past no longer exists — it is memory, a mental representation of events that no longer have substantial reality. The future does not yet exist — it is anticipation, an imaginative projection of events that may never materialise. When we say that "the past was long" or that "the future will be brief," we are speaking imprecisely. There is no extension in the past because the past is no more. There is no duration in the future because the future is not yet. The only temporal reality is the present — this instant that, in the moment we try to name it, has already become past.

But even the present, Augustine perceives, is problematic. The present has no duration. If it had duration, it would be divisible into parts: one piece still future, another already past. The pure present, the absolute now, is a point without extension, an infinitesimal boundary between what no longer is and what is not yet.

It is a knife one atom thick cutting through reality.

How can we live in what has no duration? How can we inhabit a point without extension?

Augustine's solution is to say that past, present and future exist in the soul as three modes of attention. There is the "present of past things" (memory), the "present of present things" (immediate attention) and the "present of future things" (expectation). All three are modes of present consciousness. When I remember yesterday, I am not accessing yesterday itself — I am having a present experience of recollection. When I anticipate tomorrow, I am not touching tomorrow itself — I am having a present experience of projection. The only time that exists is now, but the now can contain representations of non-nows.

This analysis is brilliant. It is also a precise description of the temporal trap in which *Homo sapiens* is caught. Augustine correctly identified that only the present of the present is real. But Augustine himself spent his life tormented by the past and obsessed with the future. The *Confessions* are fundamentally an exercise in rumination — the sins of youth, the late conversion, the years wasted in manichaean heresy, the guilt of concubinage, the anguish over his mother's death. Augustine is not present in his *Confessions*. He inhabits linguistic ghosts of events that no longer exist.

And after spending years writing about a past that does not exist, Augustine dedicates the final years of his life to *The City of God* — twenty-two books of eschatology, elaborate speculation about a future that does not yet exist. He constructs a vast narrative about two cities, the earthly and the heavenly, about the final judgement, about eternal destinies, about future redemption. All of this is linguistic projection about events that may never happen. Augustine knows, intellectually, that only the present is real. But he lacks the cognitive machinery to live that truth. This is the tragic condition of a species whose language allows it to inhabit times that do not exist.

Homo erectus did not have this problem. Not because it was wiser, but because it lacked the cognitive architecture necessary to create the problem. Erectus lived in the present naturally. Without a recursive language capable of deep temporal displacement, it remained anchored in immediate reality. The distinction here is not between temporal awareness and its absence — functional anticipation of near events is documented across vertebrates and was certainly present in erectus — but between that functional anticipation and the unlimited recursive displacement that only full propositional language enables: the capacity to inhabit, as psychological present, events decades removed and purely hypothetical. It could remember recent events and act on them. It could anticipate near events and prepare for them. But it lacked the capacity to psychologically inhabit remote times, to relive traumas from decades ago, to project detailed scenarios of distant futures, to construct autobiographical narratives turning life into a story with imagined beginning, middle and end.

Erectus could react to present death — the companion who just died, the predator threatening to kill now. Without the neuroanatomical substrate for recursive self-referential language — absent in erectus endocasts — there is no plausible cognitive mechanism for constructing "my own inevitable future death" as an abstract proposition detached from a concrete threat. The terror, if any, would have been situational, not existential.. It did not spend sleepless nights contemplating an annihilation decades away. It lacked the linguistic capacity to formulate "I will die" as an abstract proposition detached from a concrete threat. Death was an occasional event, not a permanent existential horizon.

Augustine formulates the truth and then spends his life violating it. He knows the present is the only reality, but inhabits memories and anticipations. He recognizes that time exists only as a mode of present consciousness, but dedicates present consciousness to constructing elaborate representations of pasts and futures. It is like someone who discovers they are in a prison, describes the bars with precision, and then spends the rest of their life decorating the walls of the cell instead of trying to escape. Except that, for sapiens, there is no escape. Language is the prison, and we cannot think without it.

The past does not exist. This is not a controversial metaphysical assertion. It is a factual description. The events we call "past" are no longer happening. The physical configurations of the Universe corresponding to those events have been replaced by different configurations. The past is not stored somewhere awaiting access. There is no cosmic archive where everything that has ever happened remains preserved. The past left traces — physical records, documents, neurological memories — but the events themselves ceased to exist the moment they happened.

But sapiens, equipped with a language permitting the construction of complex narratives about remote events, transforms these traces into psychological ghosts that inhabit the present. Memories are not objective recordings of past events. They are reconstructions, reinterpretations, fictions based loosely on neurological traces that have been modified each time they were accessed. Every time you "remember" something, you rewrite the memory. The act of recalling alters the object recalled. And yet we treat memories as reliable records, build identities upon them, suffer for them as though the events were happening now.

Trauma is the clearest example of how the non-existent past becomes psychologically present through linguistic mediation. There is a traumatic event — abuse, violence, accident, loss. The event happens and ends. In physical terms, it has ceased to exist. But for sapiens with elaborate autobiographical narrative capacity, the event does not end. It is relived repeatedly through detailed episodic memory. It is retold in words, transformed into story, incorporated into narrative identity as "what happened to me and defines me." Trauma becomes a linguistic ghost that haunts the present.

Homo erectus certainly experienced aversive events — attacks, injuries, losses of group members. It had memory of these events and probably modified its behaviour based on them. An erectus attacked by a predator at a certain location would avoid that location in the future. But it lacked the capacity to transform the event into a narrative identity, to construct the "traumatised self" as a permanent psychological category, to relive the event with vivid sensory detail decades later through internal language. The event happened, left a behavioural trace, and life continued. Erectus certainly had fear conditioning and stress responses mediated by amygdala-hippocampal circuits conserved across mammals. What it lacked was the capacity to transform an aversive event into a linguistically mediated narrative identity — "the traumatised self" as a permanent symbolic category. The specifically human elaboration of trauma — its rumination, its incorporation into autobiographical identity, its resistance to temporal closure — requires precisely that layer of recursive self-narration that erectus endocasts give us no reason to attribute to the species.

Guilt about the past is another distinctively human ghost. Not functional guilt — the immediate remorse after harming someone, which leads to repair and reconciliation. That form of guilt exists in many social mammals and certainly existed in erectus. It is the abstract guilt about remote transgressions, kept alive by internal narrative, that is specifically human. Augustine, in the *Confessions*, laments sins committed decades before. He relives them linguistically, examines them from every angle, constructs elaborate theological interpretations about what they mean for his eternal soul. The acts themselves no longer exist. The people involved may have died. The concrete consequences have long been absorbed by the flow of time. But Augustine keeps guilt alive through language, transforming past events into present torment.

Erectus did not carry decades of guilt for acts that are no longer present. It could feel immediate remorse, make concrete reparations, and move on. It lacked the elaborate language necessary to construct an internal tribunal where the past is repeatedly judged, where the sentence is never final, where the defendant is simultaneously judge and accused. This form of psychological self-torture requires linguistic capacity for autobiographical narrative, for abstract moral judgement, for maintaining symbolic representations of non-existent events as though they were present.

Identity constructed on memories is a third ghost of the past. Sapiens does not experience identity as immediate continuity. It experiences identity as narrative — a story about "who I am" constructed through the selection and interpretation of memories. This narrative is not an objective description. It is a useful fiction, constantly edited to maintain coherence, to justify present choices, to project desired futures. Inconvenient memories are suppressed or reinterpreted. Memories consistent with the desired self-image are amplified. The result is a narrative "self" bearing little correspondence to the series of actual experiences that constituted the life.

Augustine constructs an elaborate narrative identity in the *Confessions*: the sinner who converts, the intellectual who finds truth, the son who loses the beloved mother. This narrative gives meaning to his life, provides structure for interpreting experiences, creates a sense of continuity through change. But it is retroactive linguistic construction. The forty-year-old Augustine writing the *Confessions* is not the same twenty-year-old Augustine who committed the sins he now laments. Every cell has been replaced, every synaptic connection reorganised, every belief modified. The continuity is an illusion maintained by language supplying the pronoun "I" applied to successive radically different states.

Erectus had physical continuity — the same body persisting through time — and behavioural continuity — relatively stable habitual patterns. But it had no elaborately linguistically constructed autobiographical narrative about "who I am," "what defines me," "my personal journey." It did not suffer identity crises, did not ask about authenticity, did not worry about coherence between past and present. It simply was, moment to moment, without the additional layer of narrative self-interpretation that transforms existence into biography.

All these ghosts of the past — relived trauma, abstract guilt, narrative identity — are products of the linguistic capacity to construct and inhabit symbolic representations of events that no longer exist. The real past is irrecoverable. What we call "the past" is always a present construction, a useful or damaging fiction we keep alive through language. Erectus was free from these ghosts not through superior forgetting, but by lacking the language that permits creating and maintaining elaborate representations of non-presents.

The future does not exist. Again, this is not a metaphysical controversy. It is an obvious description. The events we call "future" have not yet happened. The physical configurations of the Universe corresponding to those events have not yet materialised. The future is not waiting somewhere to be discovered. It is radically indeterminate — not in the sense that we cannot predict it (though we frequently cannot), but in the more fundamental sense that there is nothing there to be predicted because it is not yet.

But sapiens, equipped with a language permitting unlimited temporal displacement, transforms future possibilities into psychological hallucinations that dominate the present. Not functional anticipations — preparing food for the next meal, seeking shelter before the rain — but elaborate projections about distant futures that may never occur. And then suffering for those projections as though they were present realities.

Chronic anxiety is the clearest manifestation of how the non-existent future becomes present torment through language. Adaptive anxiety is a response to concrete imminent threats: the approaching predator, the arriving storm, the conflict about to erupt. This form of anxiety exists in many animals and certainly existed in erectus. It is a short-term response mobilising resources to deal with a real present or immediately imminent threat. When the threat passes or is confronted, the anxiety ceases.

Chronic anxiety is radically different. It is not a response to a concrete present threat. It is a permanent state of apprehension about hypothetical futures. The anxious sapiens is not responding to a danger perceivable now. It is responding to a linguistic narrative about what might happen tomorrow, next week, next year, over the next decades. It worries about the possibility of losing its job, falling ill, being rejected, not having enough money in old age, failing to achieve abstract goals. None of these threats exists now. All are linguistic projections about possible futures.

Erectus did not have chronic anxiety about distant futures. It could have adaptive anxiety about imminent threats — the hunt that might fail, the approaching migration, the winter arriving in weeks. But these were worries tied to tangible near-future events, not abstract ruminations about hypothetical distant scenarios. It did not spend sleepless nights imagining dozens of possible catastrophes that might occur in years to come. It lacked the linguistic capacity to generate and maintain multiple simultaneous narratives about alternative futures, all competing for attention, all generating stress responses as though they were real present threats.

The terror of death is the most universal and most devastating future hallucination of sapiens. As already discussed, there is an abyssal difference between adaptive fear of a concrete lethal threat and existential terror about inevitable future annihilation. Erectus had the first. When confronted with a predator, a precipice, a fire, it felt intense fear and acted to avoid imminent death. This fear disappeared when the threat passed. There was no permanent substrate of terror about the fact that, eventually, death would come anyway.

Sapiens lives with that background terror. It knows, linguistically, that it will die. It can anticipate the event, imagine what it will be like, project itself into the moment of annihilation, contemplate the absence of experience. And it suffers for this as though death were happening now. Augustine, obsessed with eschatology, dedicates entire books to speculating about what happens after death. This obsession is not a response to a present threat — it is a response to a linguistic projection about an inevitable future event. Augustine's death does not yet exist. But he makes it psychologically present through language, and then requires elaborate narratives — resurrection, final judgement, eternal life — to make the terror tolerable.

Abstract accumulative greed is another distinctively human future hallucination. There is the functional desire for resources — obtaining sufficient food, finding adequate shelter, securing conditions for immediate survival. This desire exists in many animals and existed in erectus. It is a desire tied to concrete present or near needs. Once satisfied, it ceases. A satiated animal stops eating. An animal with adequate shelter stops seeking shelter.

Sapiens developed a qualitatively different form of desire: limitless abstract accumulation without functional ceiling. It does not accumulate only for present or near needs. It accumulates for an indefinite future, for hypothetical possibilities, for an abstract security that is never achieved. Billionaires continue accumulating wealth far beyond any conceivable concrete need. Not because they require the resources. But because language permits projecting infinite hypothetical futures where "more" always seems necessary. Accumulation ceases to be a means to a functional end and becomes an end in itself, sustained by linguistic narratives about security, success, power.

Erectus did not accumulate abstractly. It lacked the cognitive capacity to project distant abstract futures that would justify accumulation beyond functional needs. It could store food for the coming winter, maintain tools for immediate future use. But it had no abstract concepts of "wealth," "success," or "financial security" that would permit limitless functional accumulation. It lived in an extended present where needs were concrete and satisfiable.

Obsessive planning that never satisfies is the fourth ghost of the future. There is functional planning — deciding where to hunt tomorrow, which route to follow in migration, how to divide tasks within the group. This planning is tied to concrete actions in the near future. Once the plan is executed, the planning ceases. Erectus certainly had this capacity for functional planning.

Sapiens developed a qualitatively different planning: the construction of elaborate plans for distant futures, with multiple contingent scenarios, with abstract goals that are never definitively achieved. Career planning, financial planning, life planning — all involve projections across decades, all depend on assumptions about futures that may never materialise, all generate anxiety when reality inevitably diverges from the plans. And even when a plan is "completed," there is no final satisfaction. New goals are established, new plans constructed, the process continues indefinitely. Language permits planning to become an endless activity in itself, detached from any concrete functional satisfaction.

Augustine obsessively plans his future salvation. He writes about *The City of God*, about the final judgement, about eternal redemption. These are not plans for concrete near actions. They are elaborate projections about an eschatological future that may never occur. And the elaboration of these plans brings no peace — it brings more anguish, more questions, more need for theological refinement. Planning becomes an end in itself, a way of inhabiting the future instead of living the present.

All these future hallucinations — chronic anxiety, the terror of death, abstract greed, obsessive planning — are products of the linguistic capacity to construct and inhabit symbolic representations of events that do not yet exist. The real future is indeterminate. What we call "the future" is always a present construction, a projection based on extrapolations, desires and fears. Erectus was free from these hallucinations not through superior serenity, but by lacking the language permitting the creation and maintenance of elaborate representations of non-presents.

Among the many ways in which language shapes temporal experience, brazilian literature offers one of its starkest demonstrations in the work of Graciliano Ramos — one of the most rigorous prose stylists of the twentieth century. His 1938 novel *Barren Lives* presents a character whose linguistic poverty is not circumstantial but constitutive: it does not limit his life from the outside; it structures his life from within. Ramos's prose is austere, stripped of ornament, exacting in its severity — a language that exposes consciousness by subtraction rather than description. In this, he stands closer to Kafka and Beckett than to the exuberant traditions with which Latin American literature is often associated.

Fabiano, the migrant protagonist, is anatomically a complete ***Homo sapiens*** — possessing the same cranial volume, the same neural architecture, the same phonatory apparatus as any literate human. But he operates with a linguistic repertoire so restricted, so devoid of abstraction, that his experience of the world approaches the erectus condition more than the typical sapiens condition. Wittgenstein argued that "the limits of my language denote the limits of my world." Fabiano is the empirical demonstration of that proposition: his linguistic limits determine the limits of his temporal experience.

Fabiano's past does not exist as an elaborate narrative, as a semantically organized memory, as an identity constructed on stories he tells himself. He does not ruminate about wrong choices, does not mentally reconstruct alternative paths he might have taken, does not carry abstract guilt about acts that are no longer present. When the drought expels him from the farm, he simply moves — not because he weighed alternatives, not because he projected hypothetical futures, but because the hunger is concrete, the thirst immediate, the dead cattle are there before his eyes. His memory exists — he recognizes places, people, situations — but it remains predominantly episodic, imagistic, not linguistically elaborated into abstract self-referential narratives. It is a memory closer to animal memory: functional, contextual, without the additional layer of symbolic meaning that transforms the lived into personal history laden with existential significance.

Fabiano's future also does not exist as an abstract projection, as a plan, as a source of chronic anxiety about threats that have not yet materialized. He does not fear "death" as a universal category — he fears this drought, this hunger, this yellow soldier who humiliates him. When he imagines killing the yellow soldier, he does not construct a detailed plan with anticipation of legal, social, or moral consequences. The rage is visceral, immediate, and dissipates when the concrete stimulus disappears. Fabiano does not accumulate abstract wealth, does not build symbolic patrimony, does not live obsessed with a future he must secure through obsessive planning.

His relationship with time is fundamentally presentist — not by philosophical choice, not by contemplative wisdom, but by linguistic limitation. He simply does not possess the symbolic tools necessary to inhabit the future psychologically in the way the typical sapiens does. When Ramos writes that Fabiano admired the "long and difficult words" of his wife, Sinhá Vitória, he is showing precisely this: the painful awareness of an inaccessible linguistic world, a world through which other humans move with ease but which remains closed to him.

Fabiano's tragedy does not lie in cognitive inferiority — he solves complex survival problems requiring considerable practical intelligence. The tragedy lies in being a sapiens with the complete hardware but without the linguistic software that would allow him to participate fully in the temporal experience characteristic of the species. He lives in a kind of evolutionary limbo: biologically sapiens, experientially closer to erectus. And therein lies the cruel irony: Fabiano suffers less than he would if he were literate, if he mastered abstract language, if he could construct complex narratives about past and future. His linguistic poverty protects him from the "ghosts of the past" and the "hallucinations of the future" that chronically torment the typical sapiens.

He experiences suffering — hunger hurts, humiliation hurts, fear before concrete danger hurts. But he does not experience the additional suffering, specifically human, of living simultaneously in three incompatible times: a past that no longer exists but persists symbolically as guilt and identity, a future that does not yet exist but already oppresses as anxiety and existential terror, and a present that should be the only real thing but remains perpetually neglected, eclipsed by the temporal projections that language permits constructing.

Saint Augustine identified the "present of the present" as the only true temporal reality — the single instant in which existence actually occurs. Fabiano inhabits that instant. Not by virtue, not by contemplative achievement, but by the same limitation that protected the erectus: the absence of the symbolic scaffolding required to escape it. And perhaps precisely for that reason, despite all his material misery, he remains closer to the condition of a species that lived a million and a half years without needing to invent religions to fill the void created by living psychologically absent from the only time that exists.

Religion is not a historical accident or a divine revelation. It is the inevitable symptom of the temporal displacement of ***Homo sapiens***. Once a species develops recursive language capable of unlimited temporal displacement — a species that lives psychologically absent from the present, inhabiting ghosts of the past and hallucinations of the future — that species will, with remarkable cross-cultural consistency, construct belief systems attempting to make sense of the suffering generated by that displacement. The causal mechanism is not speculative: a species living psychologically outside the present requires integrative narratives to connect the disconnected temporal fragments — past guilt, future terror, present meaninglessness — into a coherent story. Religion is the most cross-culturally consistent solution to that integration problem.

All religions perform the same structural operation: they offer narratives about a mythical past explaining the present condition, and promises about a future that make present suffering tolerable. There is always a past of origins — creation, fall, initial revelation — explaining why we are here and why we suffer. There is always an eschatological future — judgement, redemption, salvation, paradise — promising to resolve present suffering through a transformative event that has not yet occurred. And there are prescriptions for the present — rituals, commandments, practices — supposedly connecting mythical past and promised future.

Augustine's christianity illustrates with unusual clarity — precisely because Augustine was unusually articulate about his own psychological processes — the structure that all major religions share. There is the mythical past of *Genesis*: perfect creation, Adam's fall introducing sin and death, the long history of patriarchs and prophets. There is the eschatological future: the second coming of Christ, the final judgement, the resurrection of the dead, eternal life in the *City of God* for the saved and eternal damnation for the lost. And there are the prescriptions for the present: faith, repentance, sacraments, moral discipline. The entire structure is an attempt to give meaning to a life lived outside the present, offering explanations about a past that no longer exists and promises about a future that does not yet exist.

Erectus had no religion in this sense. It may have had rudimentary ritual behaviors — possible symbolic behaviors — though unambiguous evidence of intentional burial is absent from the erectus record proper, and what is occasionally cited (Atapuerca, Sima de los Huesos) involves ***Homo heidelbergensis***, a related but distinct lineage — but it had no theology, no mythical narratives about origins and destinies, no doctrines of salvation or eternity. Not because it was less intelligent, but because it lacked the linguistic machinery necessary to construct such abstractions.

Religious metaphysics are attempts to fill the void created by living psychologically absent from the present. When you live in the immediate present, you need no explanations about where you came from or where you are going. You are simply here, now, responding to what is. But when you live linguistically inhabiting pasts and futures, when the present becomes a narrow corridor between ghosts and hallucinations, you need narratives connecting these phantasmatic times into a coherent story. Religion provides these narratives.

Eschatology — the doctrine of last things — is particularly revealing. All major religions have eschatology: Hindu cycles of creation and destruction, Buddhist nirvana as liberation from rebirth, the Jewish messianic era, the Christian second coming and final judgement, the Islamic day of resurrection, Zoroastrianism's final battle between good and evil. These narratives are variations on the same theme: the promise of a transformative future that will resolve present suffering.

Augustine's *City of God* is a monumental eschatological construction. He describes the conflict between the earthly city (governed by love of self to the contempt of god) and the heavenly city (governed by love of god to the contempt of self). He projects a future in which this conflict will be resolved, the just vindicated, the wicked punished, and history brought to a definitive end. All of this is linguistic projection about a future that does not exist. It serves to make present suffering — the sack of Rome, the decadence of the empire, human mortality — tolerable by inserting it into a narrative promising future resolution.

Soteriology — the doctrine of salvation — is equally symptomatic. All religions offer salvation: hindu moksha, buddhist nirvana, the jewish world to come, christian eternal life, islamic paradise. Salvation is always future, always conditional on present behavior, always promised but never verifiable. It is the supreme future hallucination: the promise that present suffering will be resolved at an indefinite future time through supernatural means.

Erectus needed no salvation because it had no linguistic concept of damnation. It did not live with existential terror of inevitable future death, so it needed no promises of eternal life. It carried no abstract guilt about the past, so it needed no narratives of redemption. It suffered no anguish about the meaning of life, so it needed no theodicies explaining suffering. It simply existed, confronting the concrete challenges of the present with the resources available in the present.

The attempt to return to the present through religious practices is the final irony. Many spiritual traditions recognize that living in the present is desirable. Buddhist meditation emphasizes mindful attention to the present moment. Christian mysticism speaks of divine presence experienced in the now. Sufi practice cultivates awareness of the instant. But all these practices are linguistically formulated, transmitted through language, and monitored through linguistic self-observation. "Am I present?" is a linguistic question that immediately removes you from the present. "I should stop thinking" is a thought that perpetuates thinking.

Erectus was present naturally. Not as a spiritual achievement, not as the result of contemplative practice, not as an altered state of consciousness. It was simply the normal condition of an organism without recursive language to displace it temporally. It needed no techniques to "return" to the present because it had never left it. Presence was the default state, not a goal.

Augustine knows that only the present of the present is real. He writes eloquently about this. But he cannot live it. He spends his life inhabiting a past that no longer exists (*Confessions*) and a future that does not yet exist (*City of God*). The religion he helps construct is an attempt to make sense of a life lived outside the present, offering narratives connecting mythical past and eschatological future through prescriptions for the present. But the prescriptions do not lead to the present. They lead to more narratives, more theological elaboration, more temporal displacement.

Sapiens developed a double life: biologically in the present — the body here and now, processing sensations — but psychologically in the past and future — the mind inhabiting memories, anticipations, projections. This dissociation between biological presence and psychological absence is the primary source of human suffering. We are not where we are. We are always somewhere else linguistically constructed.

Erectus had no such double life. It was where it was, when it was. Biology and psychology coincided in the present. There was no additional layer of linguistic representation creating a parallel symbolic world inhabited by the mind while the body occupied the physical world. There was only one life, lived in the only time that exists.

Religions promise to resolve the suffering created by temporal displacement through more temporal displacement — promises of celestial futures, explanations of mythical pasts. But the real resolution would be to stop linguistically inhabiting times that do not exist and simply be here, now. That is precisely the capacity we lost when we developed recursive language. We gained the ability to construct elaborate theological cathedrals attempting to make sense of our absence from the present. We lost the ability to simply be present, as erectus was for a million and a half years.

V.
Success and its shadows

No themes are so human as those that reflect for us, out of the confusion of life, the close connexion of bliss and bale, of the things that help with the things that hurt, so dangling before us for ever that bright hard medal, of so strange an alloy, one face of which is somebody's right and ease and the other somebody's pain and wrong.

HENRY JAMES
What Maisie knew

What Maisie knew tells the story of a child divided between divorced parents who use her as a pawn — each extracting advantage while she absorbs the consequences. James's two-faced medal is the perfect metaphor for what we call "civilization": the obscene fortune of a few as the inseparable, mirrored face of the misfortune of many. These are not distinct phenomena accidentally connected; they are the same coin, minted in the same act. The "strange metallic alloy" James evokes is the very substance of sapiens social organization — an alloy that can exist only through symbolic abstractions that allow one person to accumulate while another perishes, that let one side of the medal gleam precisely because the other remains in shadow.

Here we finally reach solid ground. We are no longer dismantling cosmic illusions or describing the cerebral anatomy of sapiens suffering. We are establishing the fundamental criterion that allows us to judge which species truly won the evolutionary contest — remembering, of course, that nature itself recognizes no race toward anything, which makes the metaphor all the more revealing of our bias — and the answer will be as mathematically incontestable as it is philosophically devastating.

The conventional metric of evolutionary success that permeates almost every discussion about humanity is anthropocentric from head to toe. We measure "progress" by technology, art, philosophy, science, social complexity. We measure "superiority" by the capacity to transform environments, create symbols, manipulate abstractions. We implicitly assume that the species that does more, changes more, and dominates more is the winning species. This metric is so deeply rooted in our self-image that we rarely pause to question its validity. But from nature's standpoint — which cares nothing for our symphonies, theorems, or cathedrals — this metric is utterly irrelevant.

Ernst Mayr, along with a broad consensus in neo-darwinian population biology, argued that species longevity is among the most reliable indicators of adaptive adequacy. Nature appears to measure evolutionary success in an overwhelmingly simple way: how long a species manages to keep its lineage functioning in integrated fashion with the ecosystem on which it depends. No matter how impressive a species' technology, if that technology accelerates its extinction. No matter how sophisticated its Art, if that Art coincides with the collapse of the conditions that permit its survival. The only objective criterion of evolutionary success is sustainable longevity — and by that criterion, ***Homo erectus*** surpasses ***Homo sapiens*** by a factor of at least six to one, a geological verdict no technology has yet overturned.

The distinction between efficacy and effectiveness is central to understanding why the hypertrophied brain of ***Homo sapiens*** represents, paradoxically, both an adaptive triumph and an evolutionary disaster. Individual efficacy refers to an organism's capacity to solve complex problems, innovate, manipulate the environment, and create tools and strategies that increase its chances of immediate survival and reproduction. Collective effectiveness refers to the long-term sustainability of the species as a whole — its ability to keep the lineage functioning in integrated fashion with the ecosystem on which it depends, without precipitating the collapse of the systems that sustain it.

Homo sapiens is extraordinarily efficacious. No other species has developed an equivalent capacity for solving abstract problems, creating complex technologies, or transmitting cumulative knowledge across generations through symbolic culture. Sapiens can anticipate future threats and devise preventive strategies; can cooperate on massive scales through shared fictions — nations, religions, corporations; can manipulate the environment in ways no other organism has ever managed. This individual and group efficacy is undeniable: it is what allows a small bipedal mammal of sixty to eighty kilograms to transform the planet, build cities of tens of millions, send machines beyond the Solar System, decipher the genetic code, and attempt to manipulate evolution itself.

But extraordinary individual efficacy does not guarantee collective effectiveness. Quite the contrary: in the case of sapiens, individual efficacy tends — at civilizational scale — to operate at the expense of collective effectiveness. Not by necessity, but by the structural logic of a brain optimized for proximate problem-solving in a world now shaped by distal consequences. The more technology we develop, the more rapidly we exhaust non-renewable resources and destabilize planetary systems. The more symbolic abstractions we create to coordinate large-scale cooperation, the more complex and violent become the structures of domination those abstractions permit.

The reason is that sapiens' individual efficacy is the product of cognitive capacities that generate, as an inevitable by-product, a series of psychological and social pathologies that undermine collective effectiveness. The capacity to anticipate the future and recall the past — essential for complex planning — also generates chronic anxiety about threats that may never materialize and rumination about events that cannot be changed. The capacity to symbolize and abstract — essential for Language, Mathematics, Science — also generates concepts of property, debt, and hierarchy that permit structural inequality on scales impossible for species without complex language. The capacity to cooperate through shared fictions — essential for states, religions, corporations — also generates genocidal wars in the name of those same fictions, and systematic exploitation justified by narratives of racial, national, or religious superiority.

Homo erectus, by contrast, possessed sufficient individual efficacy to survive and prosper in diverse environments, but lacked the cognitive capacities that generate sapiens pathologies. Erectus made tools, controlled fire, hunted cooperatively, and evidence suggests it provided care for injured group members — behaviors requiring intelligence, empathy, and planning. But erectus had no complex symbolic language permitting radical temporal displacement or the abstraction of concepts such as property over unoccupied land or debt transmissible across generations. It had no capacity to create shared fictions of group superiority justifying the extermination of other groups. It lacked the hypertrophied default mode network and the expanded dorsolateral prefrontal connectivity that in sapiens sustain chronic temporal displacement — the neurological substrate of existential rumination.

The result of this fundamental neurological difference is that erectus achieved a sustainable equilibrium between individual efficacy and collective effectiveness. It had enough intelligence to solve adaptive problems without possessing excess intelligence capable of generating virtual adaptive problems that are potentially terminal. It lived in the material present, responding to concrete challenges as they arose, without the capacity to become lost in catastrophic future projections or nostalgic past idealizations that characterize sapiens existence. It operated in small groups where hierarchies were situational and based on demonstrated competence, not abstract accumulated power transmitted institutionally.

Sapiens, by contrast, is a classic case of individual efficacy undermining collective effectiveness. We are so efficacious at manipulating the environment that we are destroying the conditions that permit our survival. So efficacious at creating weapons that we developed arsenals capable of destroying civilization multiple times. So efficacious at extracting resources that we exhaust in centuries reserves accumulated over millions of years. So efficacious at cooperating through symbolic abstractions that we created power structures allowing a small elite to dominate billions while the planet burns.

The sapiens metric of success — Technology, Art, Science, "progress" — is a metric of individual efficacy. But from an evolutionary standpoint, what matters is not how impressive our individual or cultural achievements are, but whether we manage to maintain the lineage functioning sustainably. And by that criterion of collective effectiveness, erectus surpasses us decisively. Erectus was demonstrably more effective than sapiens — and that difference becomes more undeniable with each passing decade as we accelerate toward the ecological, climatic, and nuclear collapses that our individual efficacy made possible but our collective ineffectiveness made probable.

The word "civilization" carries an almost universally positive connotation in the sapiens lexicon. It derives from the latin *civitas* (city, community of citizens) and evokes images of refinement, moral progress, and the overcoming of barbarism. To oppose civilization is to be "primitive," "savage," "backward." To defend civilization is to defend light against darkness, order against chaos, culture against raw nature. This positive semantic charge is not accidental — it is the product of millennia of elaborate narratives crafted by the very power structures that civilization permits and requires.

But if we examine civilization not for what it claims to be, but for what it actually does, the picture changes radically. Civilization is, in descriptive and non-euphemistic terms, a system that permits and institutionalizes three central characteristics: obscene material inequality among members of the same species, organized violence to maintain that inequality, and a complex of symbolic narratives that legitimize exploitation as natural, divine, or meritorious.

Begin with inequality. In contemporary hunter-gatherer societies — imperfect but informative analogues for how erectus and pre-agricultural sapiens likely lived — material inequality is minimal. There is no private ownership of land. There is no accumulation of goods beyond what can be carried during seasonal migrations. Hierarchies are situational: the best hunter coordinates the hunt, the best healer treats the sick, the elder with the most experience advises in disputes. But these hierarchies do not translate into permanent power accumulation or control over resources others need to survive. There are no kings, no slaves, no billionaires, no hungry masses while granaries overflow.

Civilization changes this radically. Sedentary agriculture permits the accumulation of food surplus, which permits social stratification, which permits a dominant class that produces no food but controls those who do. The state emerges as an apparatus of institutionalized violence to maintain this stratification. Writing emerges, in its earliest attested forms, predominantly as a tool of accounting and administrative control — the clay tablets of Uruk record grain and livestock, not verse. Cities emerge as centers of wealth extracted from the rural peasantry. And inequality explodes to biologically obscene levels.

In the contemporary world, this inequality has reached proportions that would be comical if they were not tragic: a handful of individuals possess wealth equivalent to the bottom half of the global population combined — not as a statistic of talent or effort, but as a statistic of symbolic systems channeling the labor of millions into private accounts. A billionaire does not work a million times harder than the average worker; they control symbolic systems that funnel the labor of millions into their bank accounts.

This is not a deviation — it is the coin's other face. Civilization cannot exist without structural inequality, because civilization requires surplus extracted from primary producers — farmers, artisans, laborers — to sustain classes that produce neither food nor material goods, but control producers through violence or the threat of violence. Kings, priests, warriors, bureaucrats, executives — all these categories can exist only if there is a mechanism to extract surplus from those who produce and transfer it to those who control.

That mechanism is institutionalized violence — the second defining characteristic of civilization. States maintain armies, police, prisons, and judicial systems whose fundamental purpose is to protect property and preserve the order that benefits property owners. When hungry peasants attempt to invade aristocrats' granaries during famines, they are violently repressed. When slaves attempt to escape, they are hunted and punished. When workers attempt to organize strikes for minimum conditions, they are dispersed with batons, tear gas, and arrests. Violence is not an occasional aberration of civilization; it is its operational foundation.

And this violence must be legitimized — which brings us to the third characteristic: the complex of symbolic narratives that transform exploitation into natural order, divine will, or meritocracy. Religions proclaim that kings rule by divine right, that inequality is the will of the gods, that the rich are blessed and the poor are paying for sins of past lives. Secular ideologies proclaim that the free market distributes resources according to merit, that CEOs earn millions because they produce equivalent value, that the poor are poor for lack of effort. These narratives are fictions — but they are fictions internalized by billions, fictions that make exploitation palatable, fictions that cause victims to accept or even defend the system that exploits them.

Homo erectus had no civilization because it could not have civilization. Without complex symbolic language, erectus had no means of creating concepts of property over unoccupied land, of abstract debt, or of institutionalized hierarchy transmissible across generations. Without the capacity to symbolize absence ("this land is mine even when I am not on it") or the future ("you owe me this, to be paid in three moons"), erectus lacked the cognitive tools necessary to build structural inequality. Hierarchies existed, but they were based on present demonstrated competence: the best hunter led the hunt while hunting well; once the skill declined, so did the leadership. There were no hereditary monarchies, no land ownership, no permanently separated classes divided by a material abyss.

And without structural inequality, there was no need for institutionalized violence to maintain it. Erectus certainly knew violence — interpersonal conflict is a primate inheritance far older than any hominin. But there is a categorical difference between the violence of dominance hierarchies among social mammals and the organized, codified, institutionally sustained violence that states deploy to protect property. The former is biological; the latter is cognitive — it requires the symbolic capacity to define property, to legitimate its defense, and to recruit strangers into its enforcement. And without institutionalized violence, there was no need for complex legitimizing narratives. There were no religions proclaiming divine sanction for hierarchy, no market ideologies justifying obscene accumulation.

The result is that erectus lived in what we might describe, by structural observation rather than ideological prescription, as non-hierarchical resource sharing — not utopia, but ecological and cognitive pragmatism. Small groups, material equality enforced by the absence of means to accumulate beyond physical carrying capacity, fluid hierarchies based on situational competence, and the absence of permanent institutions of domination.

Sapiens destroyed this not because the earlier way of life was inadequate — it worked for hundreds of thousands of years for pre-agricultural sapiens and for more than a million years for erectus — but because sapiens cognitive capacities permitted the alternative: civilization. And civilization, stripped of euphemisms, is the system that transforms — to borrow the formulation introduced in the opening chapter — the fortune of the few into the inseparable face of the misfortune of many: not by accident or correctable deviation, but by structural design.

The extinction of ***Homo erectus*** remains a subject of active paleoanthropological debate. There is no definitive consensus on whether erectus was eliminated by climate change, resource competition with sapiens, hybridization dissolving the lineage, direct extermination, or some combination of these factors. The fossil record is too fragmentary for absolute conclusions. But the spatial and temporal pattern suggests that the appearance of sapiens and the disappearance of erectus are not independent events — and several lines of evidence point to a significant role for sapiens violence.

The basic pattern is stark: wherever ***Homo sapiens*** appears in the archaeological record, other human species disappear relatively quickly. ***Homo erectus*** persisted in Asia for more than a million years, surviving multiple climatic oscillations, glaciations, and radical environmental changes. Yet when sapiens arrived in Asia roughly 60,000 to 80,000 years ago, erectus vanished on a geologically abrupt timescale. The last known erectus populations — such as ***Homo erectus soloensis*** in Java, currently dated between 108,000 and 117,000 years ago — overlap temporally with, or slightly precede, the regional arrival of sapiens. A similar pattern appears with Neanderthals in Europe, who coexisted with sapiens for perhaps 5,000 to 10,000 years before disappearing around 40,000 years ago. ***Homo floresiensis*** disappears around 50,000 years ago in a transition still debated, and the Denisovans — known primarily from DNA — also vanish after contact with sapiens, leaving only a genetic trace in melanesian and aboriginal australian populations.

Several hypotheses attempt to explain this pattern. Climatic shifts may have weakened erectus populations, while sapiens survived through greater behavioral flexibility. Resource competition may have displaced erectus from shared ecological niches, leading to starvation or reproductive stress. Hybridization may have dissolved erectus into the sapiens gene pool, though current genomic evidence shows little or no erectus contribution — a result that may reflect genuine absence or simply the limits of DNA preservation in tropical contexts.

But there is also the possibility of organized extermination. Direct evidence for this is nearly impossible to obtain: marks of violence on fossilized bones are rare and difficult to interpret even in recent contexts, and distinguishing intraspecific from interspecific conflict in a fragmentary record spanning hundreds of thousands of years is nearly unworkable. Yet the circumstantial evidence is difficult to ignore. Sapiens is extraordinarily violent toward other sapiens, as archaeological sites attest: mass killings in the Neolithic, skeletal remains showing trauma consistent with organized conflict, and a long historical record of genocidal violence across civilizations. Nothing in the behavioral repertoire of sapiens suggests that Paleolithic groups encountering a visibly different hominin would have behaved with greater restraint than modern sapiens displays toward its own species.

Moreover, sapiens systematically eliminates competitors for ecological space, whether through direct hunting, habitat destruction, or resource monopolization. This pattern is consistent across continents and millennia. There is no principled reason to assume that erectus — a large-bodied primate occupying overlapping niches — would have been exempt from this competitive pressure.

The cognitive capacities distinguishing sapiens also make organized extermination possible. Complex symbolic language allows the creation of narratives of group belonging and the dehumanization of outsiders. These capacities enable forms of collective violence that are psychologically impossible for species lacking symbolic abstraction. Erectus had no cognitive tools to construct the category of "them" at the level required for systematic elimination. Sapiens did — and used it repeatedly. A visibly different hominin, with distinct cranial anatomy, posture, and possibly vocalization, would have been an ideal target for such symbolic exclusion.

The rapidity of replacement further strengthens this inference. When two species compete for resources without direct violence, the expected pattern is prolonged coexistence with gradual displacement over tens or hundreds of thousands of years. What we see with erectus and sapiens is replacement on the scale of a few thousand years — a tempo more consistent with active violence, though severe reproductive disadvantage compounded by disease and resource displacement could, in principle, produce similar timescales. The point is not that extermination is the only explanation, but that it aligns closely with everything else we know about sapiens behavior toward competitors.

None of this constitutes definitive proof. It is inference from circumstantial patterns. Climate change may have weakened erectus populations; resource competition certainly occurred; diseases transmitted by sapiens may have devastated groups with no immunity. The most plausible scenario is a combination of factors — but including organized violence in that combination is entirely consistent with sapiens behavior in every subsequent context.

And if extermination played a role — if sapiens eliminated erectus not through adaptive superiority but through the uniquely sapiens capacity for abstract hatred and symbolically coordinated collective violence — then the first great event of sapiens history was not an adaptive triumph. It was a crime. The first genocide. The harbinger of all that would follow. Not the victory of the "fittest" species, but of the species capable of systematically killing competitors through symbolic coordination enabling dehumanization and violence on a scale impossible for species without complex language.

Homo erectus did not lose through evolutionary inadequacy. It had survived 1.5 million years, crossed glaciations, and migrated across continents. It lost because a competitor arrived capable not only of competing for resources, but of generating abstract hatred, collectively organized violence, and narratives transforming extermination into duty, destiny, or divine will. It lost because sapiens is not merely more intelligent — it is more violent in ways that intelligence permits and amplifies.

If we accept — even as possibility rather than certainty — that sapiens played an active role in the disappearance of *Homo erectus* through violence, we are forced to acknowledge something profoundly disturbing about our species: the first significant act of sapiens as it spread across the world was not the construction of glorious civilizations, not the creation of Art, not the development of Philosophy. It was the extermination of other human species that had as much right to exist as we do. And that first extermination established the pattern that would repeat incessantly throughout all subsequent history.

The sapiens trajectory since then is a sequence of concentric genocides. First, we exterminated other human species — erectus, neanderthals, denisovans, floresiensis, all divergent lineages sharing a common ancestor with us. Then we exterminated the Pleistocene megafauna: mammoths, mastodons, giant ground sloths, sabre-toothed cats, elephant birds, moas. The arrival of sapiens on new continents correlates systematically with the extinction of large mammals: North America 13,000 years ago, South America 10,000 years ago, Australia 45,000 years ago, Madagascar 2,000 years ago. The pattern is so consistent that paleoecologists refer to it as the "blitzkrieg model": sapiens appears, megafauna collapses — a pattern robust enough across continents to constitute a working hypothesis even where debates continue over the relative weighting of climate and hunting pressure.

Then, when there were no more other human species to exterminate and the megafauna was largely eliminated, sapiens turned violence against itself with refined efficiency. Intraspecific genocide is a historical constant: tribes exterminating tribes, empires exterminating conquered peoples, nation-states exterminating ethnic minorities, ideological regimes exterminating enemy classes. The twentieth century alone witnessed the Holocaust (six million jews), the Holodomor (four to seven million ukrainians), the "Great Leap Forward" famine (estimates ranging from fifteen to forty-five million), the cambodian genocide (1.5 to two million), the rwandan genocide (800,000 in one hundred days), the Yugoslav wars, ethnic cleansing in Myanmar — and the list continues.

Parallel to human genocide, sapiens conducts a continuous ecological genocide. Current extinction rates are one hundred to one thousand times higher than pre-human background rates. We are in the midst of the sixth mass extinction, the only one caused entirely by a single species. Not by asteroid or meteoroid, not by massive volcanic eruption, not by natural climate change — but by sapiens choices: habitat destruction, pollution, overhunting, introduction of invasive species, anthropogenic climate change. A substantial fraction of current species — with projections varying widely by scenario and taxon — faces extinction risk within this century if current trajectories hold.

The pattern is clear: sapiens did not exterminate erectus as an aberration, as a unique circumstance, as an exceptional necessity. It exterminated erectus because extermination is the standard sapiens operational mode when it encounters competition for resources or space, or when destruction yields advantage. We are not a species that occasionally commits genocide in moments of moral crisis. We are a species whose history is continuous genocide on increasing scales as technology permits.

And the cognitive capacity making this possible is exactly the capacity we proclaim as our glory: symbolic language, abstract thought, cumulative culture. Language allows the creation of the "other" category and the attribution of dehumanizing characteristics — animals, demons, plagues, subhumans. Abstract thought permits planning extermination on massive scales, coordinating violence across vast populations, creating ideologies justifying genocide as necessary, just, divine. Cumulative culture permits transmitting extermination techniques across generations, refining methods, industrializing death.

If erectus could speak — not as rhetorical flourish but as an attempt to make legible what the fossil record cannot say — what story would it tell? The pattern that followed sapiens everywhere obliges us to infer the answer. It would have been a story of invaders arriving with superior weapons, superior organization, superior cruelty. A story of attempts at coexistence destroyed by sapiens violence. A story of last populations cornered and eliminated while sapiens interpreted the encounter as evolutionary progress.

We do not know. Erectus left no words, only fossilized bones and stone tools. But the pattern that followed — extermination of megafauna, incessant intraspecific genocides, mass extinction in progress — suggests that the erectus story would have been a story of sapiens violence. The first story of sapiens violence, the harbinger of all that came after.

And if we measure success not by Technology or Art or Philosophy, but by sustainable longevity and ecological integration, then erectus won so incontestably that we should erect monuments not to our sapiens glory but to its erectus memory. We should recognize that, insofar as the available evidence allows us to infer, the disappearance of erectus was not an evolutionary triumph but an evolutionary misfortune — a consequence of contingency rather than superiority. That the species which endured did so by accident, not by merit. And that 1.5 million years of erectus stability weigh more, in ecological terms, than 300,000 years of accelerating sapiens turbulence.

But sapiens does not erect monuments to the victims of its first genocide. It does not even acknowledge it was genocide. It proclaims victory, celebrates superiority, teaches children that human evolution is a story of triumphant progress. The medal Henry James described gleams on the face turned toward us — sapiens fortune. But on the opposite face, invisible or ignored, is the erectus misfortune: a species that functioned for a million and a half years eliminated by a competitor that, in a fraction of that time, already threatens self-destruction and drags the entire biosphere with it. This is the disturbing truth at the heart of this book: that the species which endured did so by accident, not by merit. And the world has been living with the consequences of that contingency for 300,000 years — consequences whose final cost may now be approaching

VI.
Light that blinds

They no longer know the path of peace,
and justice finds no footing beneath their steps.
They have carved for themselves crooked roads,
and all who walk them shall never find peace.
So judgment stands far from us,
and righteousness does not draw near.
We look for light — and behold, only darkness;
for the breaking of dawn — and we stumble in the night.

We grope along the walls like the blind,
we feel our way as those who have no eyes.
At noon we stagger as though it were dusk,
and among the strong we are as the dead.
We roar like bears, we moan like doves;
we wait for justice, but it does not come;
for salvation — and it remains far from us.

TRITO-ISAIAH
Old Testament (59:8-11)

Isaiah's words are not ornamental; they are diagnostic. The prophet describes a people groping blindly at noon, crying out for a salvation that never comes — a literal portrait of the sapiens condition. We have carved crooked paths through abstractions layered upon abstractions, lost within symbolic systems we can no longer navigate. Judgment has already been rendered, not by a deity but by evolutionary arithmetic: one and a half million years of erectus stability against three hundred thousand years of sapiens already flirting with self-destruction.

We roar like bears and moan like doves — the animal cries of beings who imagine themselves transcendent. We stumble at midday as though we walked in darkness because our light reveals only symbols, never the material reality that sustains life. The blindness Isaiah evokes is not spiritual; it is neurological, structural — the cognitive cost of a species trapped inside the machinery of its own fictions. And this is the crown without consolation: the species that endured is the one we have come to call the 'wrong' one. The cost of that endurance is etched into our very neurology — suffering as the inevitable by-product of a brain capable of imagining everything except the way back.

Gilles Lipovetsky coined the term hypermodernity to describe the current phase of the sapiens condition, in which the defining traits of modernity — individualism, commodification, technical rationalization, the pursuit of progress — have not been superseded but radicalized to the point of qualitative inversion. We do not inhabit postmodernity as rupture; we inhabit hypermodernity as intensification carried to paroxysm. And that paroxysm is precisely what emerges when hypertrophied prefrontal and parietal cortices operate in an environment utterly dissociated from the evolutionary fabric that shaped them.

We evolved as social primates to live in groups of roughly 150 individuals — Dunbar's proposed cognitive limit for stable personal relationships, debated in its exactness but broadly consistent with the scale of hunter-gatherer bands documented ethnographically. Our cognitive architecture evolutionarily emerged in contexts where face-to-face interaction, the reading of micro-expressions, tone, and posture were adaptive. We evolved to solve immediate, concrete problems: finding food, avoiding predators, caring for offspring, navigating small and transparent social hierarchies. Likewise, our physiological rhythms evolutionarily emerged in environments structured by natural cycles — day and night, seasons, circadian patterns synchronized with the physical world.

Hypermodernity places us in megacities of tens of millions, where we brush past thousands of strangers daily in interactions too brief to register as social. It subjects us to impersonal bureaucratic systems in which decisions about our lives are made by algorithms we cannot see, executed by institutions we will never meet, based on criteria that contradict moral intuitions forged on the Pleistocene savannah. It forces us to compete in abstract global markets where value is determined not by concrete utility but by financial speculation across multiple layers of derivatives — abstractions of abstractions of abstractions.

Lipovetsky describes a hyperconsumer society in which we consume objects not for their material properties but for symbols stacked upon symbols. We do not buy a car for transport; we buy status, identity, belonging. We do not choose clothing for thermal protection; we choose brands that signal position in a symbolic hierarchy recalibrated quarterly by the fashion industry. We do not eat food; we consume "gastronomic experiences" photographed for Instagram before ingestion, where value lies in the shared digital representation rather than in the nutrients absorbed.

This symbolic irradiation — layer upon layer of abstract meaning detached from material substrate — is a direct consequence of the linguistic and symbolic capacity that distinguishes sapiens. But carried to the hypermodern extreme, it becomes pathological. We live inside a self-fabricated matrix in which most conscious experience is mediated by symbolic representations corresponding to nothing evolutionarily relevant. The same brain that once evaluated whether fruit was ripe now evaluates whether a *LinkedIn* post will garner enough likes to influence hypothetical employers in an abstract professional network. The cortex evolved to detect immediate physical threats now incessantly monitors multiple news feeds for abstract threats — global recession, climate collapse, geopolitical tension — generating anxiety as though predators were stalking us.

Lipovetsky identifies exacerbated individual narcissism as the hallmark of hypermodernity. Each person becomes a personal brand, endlessly performing for digital audiences. Selfies are not records of moments; they are curated projections of identity. Social networks are not communication tools; they are stages on which we enact edited versions of ourselves in perpetual competition for attention, validation, symbolic recognition. And this performance is exhausting because it never ends, never satisfies, always demands more content, more engagement, more optimization of the digital persona.

Paradoxically, Lipovetsky notes the rise of presentism — the immediacy of instant gratification through streaming, delivery, swipes, algorithms anticipating desires before they fully form. But this is not the present of the present that characterized erectus. It is the anxiety of symbolic gratification that never gratifies, because symbolic gratification does not activate evolutionary reward circuits the way material satisfaction does. Eating when hungry satisfies; accumulating likes never satisfies because it corresponds to no real evolutionary need, only the abstract pursuit of status within an artificially constructed symbolic hierarchy.

Barry Schwartz documented what he called the paradox of choice — a phenomenon whose magnitude varies by context and individual, but whose structural logic remains cognitively plausible: when the option space exceeds the brain's capacity for meaningful comparison, the result is not liberation but paralysis. Extended to all domains of hypermodern life — careers, romantic partners on dating apps, political and gender identities, lifestyles, personal philosophies — the paradox of choice becomes existential paralysis. We have unprecedented freedom to choose who to be, but that freedom becomes a burden because there are no evolutionarily calibrated criteria for navigating such a vast and abstract possibility space.

Lipovetsky calls this the "era of the void" — not void through lack of stimuli, but void because the stimuli are symbolic, failing to activate evolutionary meaning systems, leaving a sense of hollowness even amid material and informational abundance. We have everything and feel nothing, because "everything" corresponds to nothing the Pleistocene brain recognizes as genuine satisfaction.

Hypermodernity is the terminal stage of symbolic irradiation — the point at which the sapiens capacity for abstraction, symbolization, and temporal displacement reaches such an extreme that it severs individuals from material reality altogether. And in that severance, suffering becomes chronic and structural, because we attempt to satisfy real evolutionary needs with symbolic substitutes that can never satisfy them.

Mental disorders are not occasional aberrations afflicting a minority of dysfunctional individuals. They constitute a global epidemic of statistically devastating proportions. The World Health Organization estimates that roughly 280 million people suffer from depression and 301 million from anxiety disorders. More than 700,000 suicides occur annually — one person every forty seconds. And these figures exclude undiagnosed or underreported cases, populations without access to mental health systems, and the vast reservoir of chronic psychological suffering that never reaches diagnostic thresholds yet steadily erodes quality of life.

The standard medical model treats mental disorders as diseases — neurological dysfunctions requiring pharmacological or therapeutic correction. Neurobiological components undeniably exist: neurotransmitter dysregulation, genetic susceptibilities, trauma-induced alterations in brain structure. But interpreting the global surge in mental suffering as a purely individual medical problem ignores a glaring systemic pattern: rates of depression, anxiety, burnout, and loneliness rise consistently in hypermodern societies, correlating with urbanization, digitalization, precarious labor, and the fragmentation of communities.

The evolutionary explanation is more parsimonious: we are not sick; we are operating in an environment radically incompatible with the cognitive architecture that evolutionarily emerged in ancestral contexts. Chronic anxiety is not a defect but an adaptive response to threat — except that the system calibrated to detect immediate predators now fires continuously in response to abstract, uncontrollable dangers: economic instability, climate change, geopolitical tension, pandemics. We cannot fight or flee a financial crisis, but the limbic system does not know this; it maintains the organism in a state of vigilance that, when prolonged, becomes pathology.

Depression, in evolutionary terms, may have functioned as a response to social defeat, loss of status, or isolation — signals that current strategies were failing and required reassessment. In small, stable ancestral groups, a temporary depressive episode could prompt behavioral change or elicit social support that resolved the underlying issue. In hypermodernity, however, the sources of defeat are systemic and inescapable: global competition no individual can win, perpetual comparison with curated digital selves, structural loneliness in megacities where no one knows their neighbors. Depression becomes chronic because the conditions triggering it cannot be resolved through individual behavioral adjustments.

Loneliness is a particularly cruel example. We are intensely social primates. In ancestral environments, social isolation was an existential threat: the isolated individual faced dramatically higher risks of predation, starvation, and disease. The brain therefore treats loneliness as physical pain — a signal of mortal danger. Hypermodernity, however, produces loneliness in crowds: we live surrounded by millions yet lack the intimate, stable interactions our social neurobiology requires. Friendships are superficial and transient, communities fragmented, extended families geographically dispersed. The result is chronic loneliness even in densely populated cities — and loneliness triggers stress responses associated, in some analyses, with mortality risks comparable to heavy smoking, a finding robust enough in meta-analytic literature to constitute a major public health concern.

The absence of purpose is another structural pathology. In ancestral environments, purpose was immediate and transparent: survive, reproduce, care for offspring, contribute to the group. Every action had visible consequence. A successful hunt fed the band; caring for a child ensured continuity; resolving a dispute preserved cohesion. Effort and meaning were directly linked. In abstract hypermodern economies, most people work in functions so specialized and distant from any tangible outcome that the connection between effort and purpose dissolves — this is alienation. The assembly-line worker produces a component without knowing the machine it serves; the financial analyst manipulates symbols without grasping their real-world impact; the call-center employee recites scripts without addressing underlying problems. Work becomes alienated in the Marxist sense and existentially empty in the existentialist sense — an activity that consumes life while producing no meaning recognizable to a Pleistocene brain.

This epidemic of suffering is intensified by an economic structure that rewards psychopathology. Corporations select executives with psychopathic traits — diminished empathy, instrumental manipulation, pursuit of power without moral restraint. Political systems elevate narcissists who perform exaggerated confidence and promise simplistic solutions to complex problems. Social media reward empty performativity, where authenticity is a competitive disadvantage. The natural compassion we inherited as social primates — empathy, reciprocity, cooperative instincts — is systematically suppressed by institutional incentives that reward zero-sum competition and punish vulnerability.

The conclusion is inexorable: the epidemic of psychological suffering is not accidental. It is not the result of increasing individual fragility or moral decline. It is the structural consequence of a species with a brain shaped on a Pleistocene savannah operating within a hypermodern matrix of abstractions layered upon abstractions. The suffering is not pathological; it is the predictable response of a Pleistocene brain confronted with hypermodern conditions it was never built to process. The problem is not in individual brains; it is in a civilization forcing Pleistocene minds into worlds their accidental evolutionary history never prepared them to inhabit.

If hypermodernity produces chronic psychological suffering at the individual level, it produces terminal existential threats at the collective level. And both emerge from the same neurological hypertrophy: the expanded prefrontal and parietal cortices enabling abstract planning, symbolic manipulation, and large-scale coordination through shared fictions — capacities conferring extraordinary individual efficacy while generating catastrophic collective ineffectiveness.

Anthropogenic climate change is the paradigmatic example. For hundreds of thousands of years, the terrestrial atmosphere maintained a relatively stable composition, permitting the development of complex ecosystems and the agriculture sustaining civilizations. But in merely two centuries since the Industrial Revolution — an evolutionary blink — sapiens altered atmospheric chemistry through fossil-fuel combustion to the point of triggering global warming now destabilizing planetary climate systems.

Pre-industrial atmospheric CO_2 concentration was roughly 280 parts per million, oscillating naturally between 180 and 300 ppm during the glacial–interglacial cycles of the last 800,000 years. In 2023, it surpassed 420 ppm — a level without precedent in at least 3 million years of Earth history. The result is increasingly extreme climate patterns: heat waves killing thousands, droughts destroying harvests, forest fires consuming entire ecosystems, intensified hurricanes, and sea-level rise threatening coastal cities where a substantial proportion of the global population lives.

The tragedy is that climate change does not occur through ignorance or technical incapacity. It occurs because the global economic structure rewards the maximization of short-term profit and penalizes long-term sustainability. Fossil-fuel corporations have known about climate change for decades — internal research from the 1970s documented the risks — yet funded disinformation campaigns because acknowledging the problem threatened their business model. Politicians know, but avoid action because the costs of energy transition are immediate and concentrated while the benefits are diffuse and future, inverting the electoral calculus.

This is a temporal-displacement problem: sapiens' capacity to project future consequences permits recognizing the climate threat, but that same capacity generates economic and political structures in which the future is systematically discounted in favor of present gains. Erectus could not cause climate change because it lacked the technology to burn fossil fuels on an industrial scale — but it also lacked the symbolic structures of abstract capital accumulation that would motivate doing so even if it could. It lived in the material present; it could not sacrifice the future of unborn generations for present profit abstracted into numbers in a bank account.

Nuclear arsenals are the second example. Sapiens developed the capacity to release nuclear energy — a triumph of Theoretical Physics and Applied Engineering — and immediately converted that capacity into weapons capable of destroying civilization multiple times over. The combined arsenals of the United States and Russia contain thousands of nuclear warheads — enough to trigger nuclear winter many times over. The use of even a fraction would inject sufficient soot into the stratosphere to block sunlight, collapse global temperatures, destroy agriculture, and kill billions through starvation even in countries not directly struck.

And nuclear arsenals do not exist by rational defensive necessity. They exist by the logic of mutually assured destruction (M.A.D.) — a concept intelligible only to a mind capable of symbolic abstractions about deterrence, balance of power, and geopolitical projection. No other animal would accumulate self-destruction capacity as a security strategy. But sapiens, operating through abstractions layered upon abstractions, built systems in which the threat of global annihilation is treated as stability.

The risk is not theoretical. Multiple incidents brought the world within minutes of nuclear war through false alarms, communication errors, and misinterpretations during crises. The 1983 Stanislav Petrov incident, in which a soviet officer chose not to escalate what instruments indicated as an incoming american strike — a decision whose full context remains partially classified but whose consequence was the absence of nuclear retaliation. The Cuban Missile Crisis of 1962, where the decision not to launch depended on specific individuals resisting pressure from subordinates and automated systems. The longer the arsenals exist, the greater the statistical probability of accidental or intentional use during a crisis.

The third threat is the sixth mass extinction. Current extinction rates are estimated at 100 to 1,000 times higher than natural background rates — meaning species are disappearing hundreds of times faster than they would without anthropogenic pressure. Since 1500, approximately 900 species have been documented as extinct, but the real number is certainly far higher because most species vanished before ever being catalogued.

Habitat destruction is the primary driver: deforestation for agriculture, urbanization, mining, infrastructure. Tropical forests — repositories of the majority of terrestrial biodiversity and major atmospheric carbon sinks — are being destroyed at alarming rates. The oceans are acidifying due to CO_2 absorption, warming due to climate change, polluted by plastics and chemicals, and depleted by industrial fishing removing biomass faster than populations can replenish.

Extinction is not merely an aesthetic or sentimental loss. Ecosystems are complex networks of interdependencies; the removal of keystone species can trigger cascading collapses. Pollinators — bees, butterflies, bats — are in sharp decline; without them, agriculture collapses because most crops depend on pollination. Top predators — wolves, large felids, sharks — regulate herbivore and prey populations; their removal triggers population explosions that devastate vegetation and alter entire ecosystems.

All these threats — climate change, nuclear arsenals, mass extinction — are products of the cognitive capacities distinguishing sapiens. Industrial technology requires abstract planning, advanced mathematics, and coordination among thousands of specialists across generations. Nuclear weapons require theoretical physics developed over centuries and engineering mobilizing the resources of entire nations. The globalized economy destroying habitats requires abstract financial systems, corporations operating across dozens of countries, and markets coordinating billions of transactions.

Erectus lacked these capacities, so it could not create these threats. It lived in equilibrium with ecosystems because it had no technology to destabilize them and no symbolic structures of accumulation that would motivate doing so. It could not build weapons of mass destruction because it had neither the theoretical science nor the social organization at the necessary scale. It could not cause mass extinction because its ecological impact was limited by simple technologies and relatively small populations.

Sapiens can do all of this — and is doing it. Extraordinary individual efficacy translates into terminal collective ineffectiveness. We are so efficacious at manipulating the environment that we are destroying the conditions permitting our survival. So efficacious at creating weapons that we developed the capacity for self-annihilation. So efficacious at extracting resources that we are provoking the collapse of biodiversity that took billions of years to emerge.

The final irony: the hypertrophied brain permits recognizing these threats with unprecedented scientific clarity. We have climate models projecting consequences of different emission scenarios. We have nuclear non-proliferation treaties documenting risks. We have ecological data quantifying extinction rates. But knowledge does not translate into action, because the economic and political structures created by that same hypertrophied brain incentivize precisely the behaviors aggravating the threats.

It is Cioran's paradox: the more conscious we become, the closer to catastrophe. Consciousness does not save; it only permits seeing with clarity the abyss toward which we run without being able to stop.

The timeline speaks with a clarity that leaves no room for illusions. ***Homo erectus***: 1.5 to 1.9 million years of sustainable existence, crossing glaciations and interglacial periods, dispersing from Africa to Asia, maintaining ecological equilibrium, living without generating structural psychological suffering or terminal existential threats. ***Homo sapiens***: 300,000 years, of which the last 10,000 under "civilization" marked by obscene inequality and recurrent genocidal violence, and the last 200 accelerating toward possible self-destruction through climate change, nuclear arsenals, and biodiversity collapse.

If evolutionary success is measured by the only non-anthropocentric metric available — long-term persistence within environments whose stresses remain within the alostatic range shaped by the evolutionary contingencies that once maintained ecological homeostasis — then erectus won so incontestably that any sapiens proclamation of superiority is the narcissistic delusion of a feverish species. But the victory of erectus was not due to technological, artistic, or philosophical superiority.

Given everything examined throughout this work, it is reasonable to infer that the victory of erectus was due precisely to the absence of those things. Erectus prevailed because it did not possess expanded prefrontal and parietal cortices generating existential anxiety, chronic depression, organized genocidal violence, abstract wealth accumulation, and the capacity to destabilize the planetary climate. It prevailed because it operated as individuals in collective survival mode, anchored in the material present, without the cognitive machinery to lose itself in phantoms of the past or hallucinations of the future.

Sapiens, by contrast, is a brilliant neurological experiment in individual efficacy but catastrophic in collective effectiveness. The hypertrophied brain permits solving abstract problems with unprecedented mastery — but also creates problems that would not exist without that same hypertrophy. It is like a tumor increasing cognitive capacity while killing the host organism: an adaptation conferring short-term advantage while guaranteeing long-term extinction.

And the extinction of erectus — the last population disappearing some 50,000 years ago — was almost certainly not the result of evolutionary inadequacy. The most plausible interpretations, given the archaeological and paleoanthropological evidence, point toward some form of sapiens-driven displacement or violence. The first significant act of the "superior" species may well have been the elimination of a competitor that had functioned effectively for more than a million years — not through ecological necessity or fair competition over resources, but through capacities for abstract hostility and symbolically organized collective aggression that erectus likely neither possessed nor could have anticipated

That inaugural violence established the pattern that would repeat incessantly: the extermination of Pleistocene megafauna, the continuous intraspecific genocides, the ongoing mass extinction. Sapiens does not commit genocide occasionally in moments of moral aberration; genocide is its standard operational mode whenever it encounters competition or the opportunity to extract benefit through destruction.

And the capacity enabling this — symbolic language, abstract thought, cumulative culture — is precisely the capacity we proclaim as our glory. But glory is a euphemism. Language permits creating the category "other" and dehumanizing. Abstract thought permits planning extermination at massive scales. Cumulative culture permits transmitting techniques of violence across generations, refining methods, industrializing death.

Lipovetsky documents the terminal stage of that trajectory: hypermodernity as maximum symbolic irradiation, where we live in abstractions upon abstractions completely dissociated from the evolutionary fabric. We consume not objects, but symbols of symbols. We interact not with people, but with curated representations. We face not concrete problems, but abstract threats generating chronic anxiety with no possibility of resolution through fight or flight.

And Cioran offers a verdict without appeal. Not as poetic lament, but as philosophical observation: for the sapiens with hypertrophied cortex generating inevitable structural suffering, birth is not a gift; it is a sentence to decades of anxiety, depression, and a search for meaning that cannot be found — because there is no cosmic meaning, only meanings we fabricate and that never fully satisfy us because we recognize them as fabrications.

Erectus likely did not face that kind of problem. As far as current evidence allows us to infer, it did not seek "meaning" because it probably lacked the cognitive architecture required to formulate such a question. It seems to have lived, reproduced, and died without carrying the kind of existential burden that emerges only in symbolically hypertrophied brains. And perhaps for that reason — or at least partly because of it — erectus managed to persist for roughly a million and a half years. Sapiens, by contrast, carries that burden and appears to be buckling under its weight in a fraction of that time. We seem to be a species that knows too much to be at ease, yet not enough to avoid the consequences of our own complexity. Intelligent enough that we cannot inhabit the present in the way erectus likely did; not intelligent enough to build systems capable of steering us away from the trajectories that now lead, slowly but structurally, toward collapse.

The conclusion that emerges is difficult to escape: by the only non-anthropocentric metric available, erectus is the lineage that endured. Sapiens appears as an evolutionary experiment producing dazzling individual efficacy but chronic collective fragility. We may well be what later generations will experience as the 'error' — the species that happened to survive — and the world has been absorbing the consequences for roughly 300,000 years, with the final costs perhaps still unfolding.

There is something particularly cruel about the sapiens condition of those who recognize everything described in this book. Some of us — paleoanthropologists, neuroscientists, philosophers, thinkers who have spent decades studying human evolution and consciousness — can see with scientific clarity what we are: primates with hypertrophied brains operating in a radically incompatible environment, generating inevitable structural suffering, accelerating toward existential threats we cannot avoid despite recognizing them. That knowledge does not liberate. It only makes suffering more conscious.

The sapiens who understands its own condition is trapped in an existential double bind. It cannot return to being erectus — cannot switch off the expanded prefrontal cortex, cannot unlearn symbolic language, cannot stop projecting the future and ruminating about the past. These capacities are neurologically inscribed, constitutive of what it means to be sapiens. There is no evolutionary reset button permitting a return to the present-of-the-present that characterized erectus.

But neither can it transcend the biological condition and achieve the purely symbolic or spiritual existence promised by some philosophies and religions. It remains trapped in a body that ages, sickens, dies. It remains dependent on biological systems that evolved in the african savanna and function poorly in hypermodern urban environments. It remains subject to anxiety, depression, loneliness — because the neurological systems generating these responses cannot be switched off through intellectual understanding.

So conscious sapiens lives the tragedy of knowing without being able to change. It knows that civilizational inequality is structural, not accidental, but cannot dissolve civilization and return to egalitarian hunter-gatherer bands. It knows that alienated work empties existence of meaning, but must work to survive in an economy offering no viable alternatives. It knows that social networks generate hollow performativity and destructive social comparison, but cannot disconnect completely without the social isolation that would be equally pathological.

It knows that climate change threatens civilizational collapse, but continues participating in the fossil-fuel economy because there is no infrastructure for an individual-scale alternative. It knows that nuclear arsenals represent an absurd existential risk, but has no power to disarm them. It knows that the sixth mass extinction is underway, but cannot stop the habitat destruction demanded by global economies.

And at the personal level, it knows that anxiety about the future is an inappropriate evolutionary response to abstract threats, but cannot stop feeling anxiety. It knows that depression may be a response to social defeat in an unsustainable hierarchical environment, but cannot dissolve hierarchies through psychological insight. It knows that the search for meaning is a symbolic fabrication attempting to fill a void with no objective content, but cannot stop searching because the search is inscribed in neural circuits.

It is a Cioranian situation perfectly crystallized: consciousness is more than a wound; it is an incurable disease. The more conscious sapiens becomes of its own condition, the more acute the perception of inescapable imprisonment. Ignorance would be a blessing, but knowledge cannot be unlearned. Erectus had the blessing of ignorance; it lived without questioning because it could not question. Sapiens questions compulsively, and each answer reveals only more layers of structural absurdity.

Some seek consolation in religious or spiritual narratives promising transcendent meaning, life after death, cosmic moral order. But the scientifically informed sapiens recognizes these narratives as fictions — useful perhaps for psychological survival, but not true in any sense corresponding to objective reality. And recognizing fiction as fiction empties its consoling power.

Others seek consolation in hedonism — maximizing sensory pleasure, accumulating experiences, "living fully" before inevitable death. But hedonism also fails because sensory pleasures adapt, saturate, and leave a greater void after ceasing. And in hypermodernity, even hedonism becomes performative, competitive, anxious — not a genuine pursuit of pleasure, but an accumulation of experiences for identity curation on social networks.

Others seek consolation in political engagement or activism — fighting for social justice, environmental sustainability, reforms that might improve the human condition. And individual efforts are admirable, often heroic. But the conscious sapiens recognizes that the problems are structural, not solvable through incremental adjustments. Inequality is not a malfunction of civilization; it is one of its organizing principles.

Environmental destruction is not a market failure that regulation can correct — it is the inevitable consequence of economies requiring infinite growth on a finite planet. Genocidal violence is not the aberration of specific regimes — it is a recurring pattern of a species capable of abstract hatred.

So what remains? For the sapiens that knows too much — the one who recognizes neurological and structural imprisonment without exit, who sees trajectories of possible or probable collapse without the capacity to divert them — what response exists that is not despair or paralyzing cynicism? The old distinction between those who do not know, those who know and collapse, and those who know and must still choose how to stand in the face of the void returns here with renewed force. And yet, for a species that cannot unknow what it knows, even the notion of a "posture" may itself be an illusion.

Perhaps none. Perhaps the most honest conclusion is that there is no consolation, no redemption, no solution waiting beyond the horizon. Only the recognition that the lineage best adapted to long-term persistence is gone, and that sapiens — brilliant in individual efficacy yet fragile in collective outcomes — carries neurological burdens that incline it toward suffering and structural trajectories that tend toward collapse.

And perhaps the only intellectually honest gesture that remains is to look back at erectus — silent for tens of millennia, extinguished by our violence or our indifference — and acknowledge what its endurance suggests: a life anchored in the present, organized in small and relatively egalitarian groups, in equilibrium with the surrounding world, sustained for more than a million years without generating the curses that now define us.

We, sapiens, with our symphonies and theorems and cathedrals and philosophies, may be an evolutionary paradox unfolding in slow motion — dazzling in complexity, yet unable to stabilize the very systems we create. And the Cosmos remains indifferent, as it has been to the disappearance of nearly every species that ever lived, operating through processes so vast that our anguish dissolves into irrelevance.

Isaiah growled while waiting for a salvation that never arrived. We growl as well — with greater scientific sophistication, but with the same result. Salvation is distant not because it is delayed, but because it does not exist. What exists is an evolutionary trajectory that produced a brain capable of perceiving its own tragedy without possessing the means to escape it.

And perhaps that — the lucid recognition without consolation — is all that remains. Not hope, not solutions, not redemption. Only the truth that we are a species that knows too much, and that this knowledge is itself our final condemnation.

EPILOGUE

It was quite clear to me that the religious paradise of youth, now lost, was a first attempt to free myself from the "merely personal" —from an existence dominated by wishes, hopes, and primitive desires. Out there was this vast world, existing independently of us human beings, an enormous and eternal enigma, at least partially accessible to our reason.

I understood that the contemplation of this world was itself a new form of liberation [...] The possibility of grasping this impersonal world through rational means became for me, consciously or unconsciously, the supreme aim [...] Perhaps the path to this paradise was not as comfortable or as certain as the path to the religious one; but it proved trustworthy, and I have never regretted the choice.

ALBERT EINSTEIN

This is not progress. It is a catastrophe unfolding in slow motion. And the catastrophe did not begin with industrialization, nor with agriculture, nor even with the mastery of fire. It began when a primate developed disproportionate prefrontal and parietal cortices and began inhabiting not the world, but representations of the world. When it ceased to live in the present and began to dwell in phantoms of the past and projections of the future. When it traded effectiveness for efficacy, integration for domination, presence for symbolization. The catastrophe began with us, with Homo sapiens — and Einstein sensed this when he sought refuge in the "impersonal world," that territory preceding us, exceeding us, and destined to continue long after we extinguish ourselves through our own symbolic excess.

Homo erectus inhabited that impersonal world for a million and a half years. Not because it was inferior, but because it was sufficient. Its brain did not generate the illusion of separation between observer and observed, between subject and world, between lived present and imagined future. It did not construct narratives about itself because there was no hypertrophied "self" demanding constant symbolic validation. It lived in the direct contemplation Einstein described — but without needing to reconquer it through reason, as the physicist had to do after losing the religious paradise of his youth. Erectus never lost that paradise because it never created the illusion of being apart from it.

We, sapiens, created religions to console ourselves for the anguish our own brains generate. Created Science to dominate a Nature from which we imagined ourselves separate. Created Philosophy to question a meaning that only became problematic because our expanded cortices tore us from immediate experience. Created Art to express a subjectivity that is, itself, a neurocognitive artifact. All of this is magnificent, yes. But it is also pathological — the restless labor of a species that developed an organ whose main function seems to be generating problems that that same organ then attempts, vainly, to solve.

Einstein found in the impersonal world of Physics a form of liberation from the "merely personal." But *Homo erectus* never needed liberation because it was never imprisoned. It did not carry the burden of reflexive self-consciousness, of guilt stretched across time, of ambition that never satisfies, of fear of a death that exists only as an abstraction. It lived and died without transforming existence into an existential problem. And for that it endured. For that it was effective where we are merely efficacious.

The irony — quiet, but brutal — is that I am writing this. A sapiens too conscious to console itself with illusions, yet sapiens enough to need to write about it: to symbolize the very critique of symbolization. Cioran would call this the "inconvenience of having been born." I was born into the wrong species, at the wrong moment of evolutionary history, with the wrong brain for the task of simply living. I look at *Homo erectus* with the same melancholic envy with which Einstein looked at the impersonal world that existed before him and will continue after us. But while Einstein could take refuge in equations, I can only take refuge in the recognition that there is no refuge.

I did not write this book to offer solutions. Solutions belong to those who still believe that a species which likely contributed decisively to the disappearance of its hominid cousins, accelerated the extinction of megafauna wherever it expanded, poisoned oceans, and altered the planetary climate in just two centuries of industry might somehow "fix" its own nature. What we call a defect is simply the way we were shaped. *Homo sapiens* behaves as it was selected to behave: maximizing short-term advantages, exhausting niches, competing with individual brilliance at the cost of collective longevity.

Nor did I write to lament. Lament presupposes that things could have been otherwise. But evolution knows no "otherwise." It knows only what occurred, through blind contingencies indifferent to our values, our moral judgments, our hierarchies of merit. Nature did not choose wrongly because Nature does not choose. It selects what persists under given conditions. And we persisted — for now — not because we were better, but because we were more aggressively adaptive at a particular moment of the Pleistocene.

What I wrote was a recognition. Cold, perhaps. Without hope, certainly. But not without a certain clarity. There is a strange dignity in acknowledging that we are an evolutionary anomaly, a neurological experiment maladapted for long-term stability. There is honesty in admitting that ***Homo erectus***, with its unchanging tools, its modest brain, its absence of symbolic art, lived more successfully than we ever will — if success is measured by the only pattern Nature seems to reveal: duration.

A million and a half years. Let that resonate. While these pages were being written, wars continued, markets oscillated, politicians dissembled, corporations devoured, individuals suffered from anxiety, depression, existential emptiness — all symptoms of cortices that evolved too fast and too far from the environments that shaped them. And nowhere, in any of these manifestations of sapiens distress, is there the faintest sign that we are approaching the longevity of erectus. Three hundred thousand years. That is all we have. And look what we have done with them.

There is no cheap optimism here. I will not say that "there is still time," because changing would require becoming something other than what we are, and that lies beyond our biological reach. I will not say that "awareness is the first step toward a solution," because awareness is part of the problem — more symbolization, more metacognition, more weight on already overburdened cortices. I will not say that "we can learn from erectus," because learning, for us, means absorbing symbols — and it was precisely the excess of symbolization that distanced us from the direct effectiveness erectus embodied.

What remains is the dignity of lucidity. Living without illusions does not save us, but it frees us from the need to invent consolations. Recognizing that our survival was an evolutionary accident — a kind of abortive outcome of contingency rather than design — does not make us wiser, but it makes us more honest. And honesty, at this late hour of the sapiens trajectory, may be the last intellectual luxury still available.

Einstein sought the impersonal world as refuge from the "merely personal." What he did not say — and perhaps could not — is that this impersonal world was already inhabited: by every form of life that never imagined itself the center of its own experience; by ***Homo erectus***, for longer than we can conceive; by the vast multitude of species that live and die without transforming existence into a metaphysical burden.

We will not return to that world. We cannot. Our brains are evolutionary enclosures from which there is no biological escape. But we can, at least, recognize the bars. We can name the jailers — hypertrophied prefrontal cortex, expanded parietal areas, uncontrolled symbolic capacity. We can understand that what we call "superior intelligence" is, in natural terms, an elaborate misalignment with long-term survival.

And we can, perhaps, live with that awareness without letting it crush us. Not because it will save us — it will not. Not because it will make us better — it will not. But because there is a peculiar freedom in recognizing one's own contingency, one's own anomaly, one's own evolutionary misfire. A freedom *Homo erectus* never needed, because it was never imprisoned — but which we, late sapiens, can glimpse, briefly, between the bars of our own minds.

This book ends, then, not with hope, but with recognition. *Homo erectus* endured by the criteria that matter. We, *Homo sapiens*, faltered before we understood the Nature of the contest. And what remains, at this final stretch of our brief passage, is the quiet dignity of acknowledging the defeat without inventing consolations.

The catastrophe continues, in slow motion. And we — the only animals capable of filming our own extinction — continue symbolizing until the end. Without optimism. Without solutions. Without illusions. Only the cold recognition that our survival was a contingency, not an achievement.

And we know it.

WHAT REMAINS

Unconditional affection operates as an ontologically irreducible phenomenon — it does not justify existence, does not confer teleological meaning, but makes it bearable to inhabit the awareness of tragedy, with an integrated acceptance of the absolute present that receives another body. It is phenomenological suspension within the tragedy, not its transcendence. The shared presence in which symbolic irradiation temporarily recedes. When you look at your son and he meets your gaze with sweetness and innocence — even knowing that this innocence will be consumed by recursive language, that he too will be enclosed — there is a moment that simply is, without demanding to signify...

BRAIN CODEX

NEUROBIOLOGIA

www.ingramcontent.com/pod-product-compliance
Lightning Source LLC
Chambersburg PA
CBHW041302210926
51847CB00010B/119
9786598812669